Leaving Certificate

New Senior BIOLOGY

Workbook

Siobhán Scott BSc, HDE, FIBiolI

Kevin Maume MA, MSc, HDE, EurProBiol, MIBiolI

Folens

Editor
Antoinette Walker

Design & Layout
Oisín Burke

Cover Design
Melanie Gradtke

Artwork
Michael H. Phillips

ISBN: 1 84131 485 4

Published in Ireland by Folens Publishers, Hibernian Industrial Estate, Greenhills Road, Tallaght, Dublin 24

Contents

Introduction

STUDYING BIOLOGY

As with all of your subjects in the Leaving Certificate, it is important that you study in the most effective way possible for **you**. There is a large amount of material to cover in biology and there are a large number of facts / terms / definitions, etc. that you will need to learn. To do this you must develop a **good study technique**. It is vital that this is developed early in your study of the subject so you do not find yourself with too much work to cover in the last year of your course.

You should develop a **regular study habit**. You should try to study at the same time, in the same place every day. This will mean that you will settle down to productive study quickly and efficiently. If study is to be most effective it should be a regular and relatively brief activity without cramming for exams. It should be done in blocks of about two or two-and-a-half hours. Within this you should study in blocks of 20 to 40 minutes, with the last five or 10 minutes used to review what you have done in the previous 15 or 30. Remember study is most productive when you undertake it during periods of maximum concentration. It is important that you realise that **time spent on study does not necessarily equal work done!**

When you are studying you should always have a pen (or pencil) in your hand and you should summarise your work as you go along. This summarising can take many forms:

- You could write short notes on cards (to minimise what you write).
- You could in some instances draw labelled diagrams, such as looking at the cell.
- You could make up rhymes and phrases to remember material e.g. **P**eas **M**ake **A**wful **T**arts (for the stages of Mitosis).
- You could make up mind maps (this requires you to draw a diagram showing the connections between the different parts of the topic (see Fig. (i)).

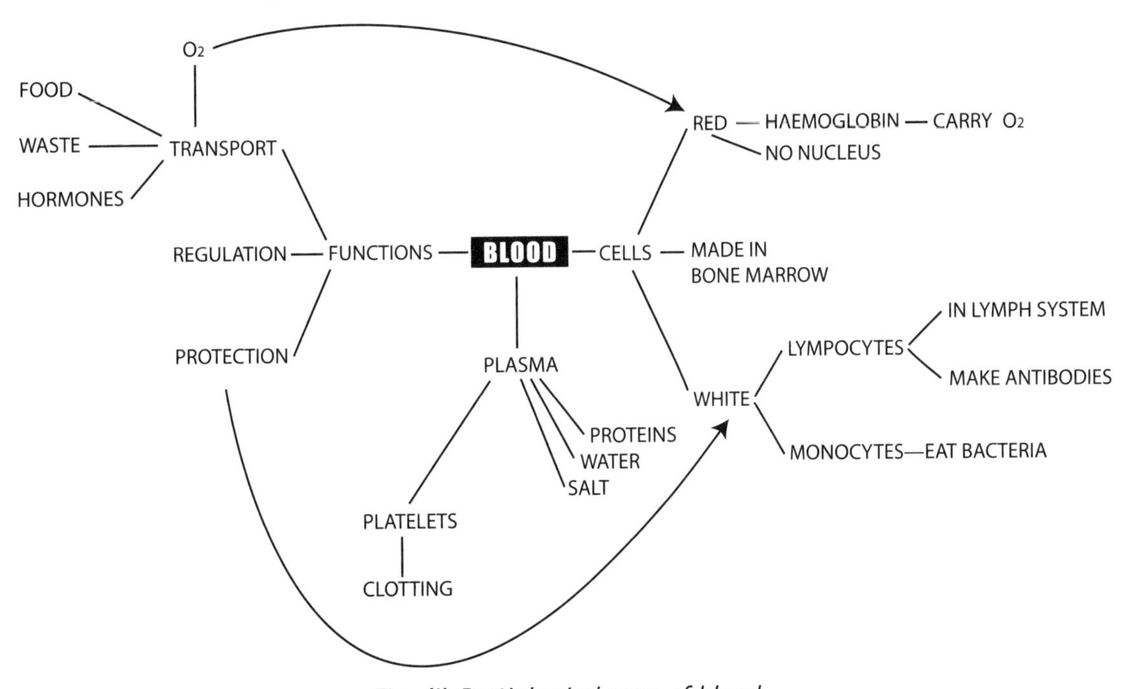

Fig. (i) Partial mind map of blood

Before you start studying a topic you should set yourself a goal, e.g. I will produce a mind map or a question to answer, e.g. 'What affects the rate of photosynthesis?'. This will focus your study and make it more productive. The summaries you produce will be your first port of call when you come to revise the topic. This will make your revision easier and shorter as you will not have to read through the text again (unless you find in a particular topic that your notes are not good enough).

You will also find that the very act of producing summaries will increase your understanding of the topic as you pick out the important concepts from the material being studied. It is important to remember that this process should be an active one where you are thinking about the material and deciding what is important not just transcribing sections of the text.

Remember to reward your self when you have **achieved** the goals you set yourself at the beginning of your study period – watch that programme on the TV, attend the party at the end of the week, etc. You should get up and move around / stretch when you have finished a topic during your two hours of study.

Check your progress regularly as you study to make sure you are learning the material:

- Try to answer questions from the textbook or workbook.
- Attempt to answer exam questions (in the same time as allowed in the exam).
- See if you can reproduce your mind maps or summary sheets.

REVISION

It is important for your success in exams that you learn to revise effectively. Much research has been undertaken on this subject and these studies have led to a number of key recommendations that you need to consider when revising.

1. How to organise your revision.
2. Length of time spent revising.
3. When revision should occur.
4. How often revision should be undertaken.

- You should make a revision timetable well in advance of your exams and you should stick to this timetable to the best of your ability.
- Pay particular attention to the areas you have difficulty understanding (it is easy revising what you already know).
- It can be useful to work with someone else. They can teach you the topics they understand better and you do the same for them. You will remember much more of any topic that you have to teach someone else!
- You have to divide your time between learning material and practising exam questions.
- Research has shown that if you study for a fixed period of time you remember more from the beginning and at the end of that period. This means that if you divide up your two-hour study period into shorter periods with short 'stretch' breaks every 30 minutes or so then your productivity will be much greater as shown in Fig. (ii).
- The frequency of your revision is very important. It has been demonstrated that you will quickly forget most of the material you read if you study it only once. This is shown in Fig. (iii). However studies also show that planned revision can greatly improve recall. If you revise after 10 minutes, your recall will improve and will develop even futher if you review the material after one day and one week, see Fig. (iv).

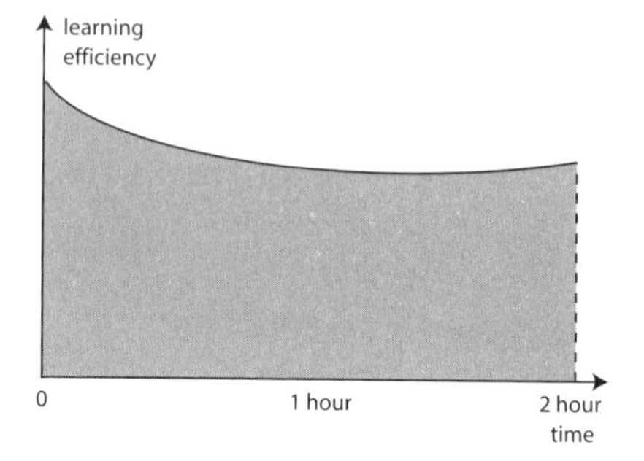

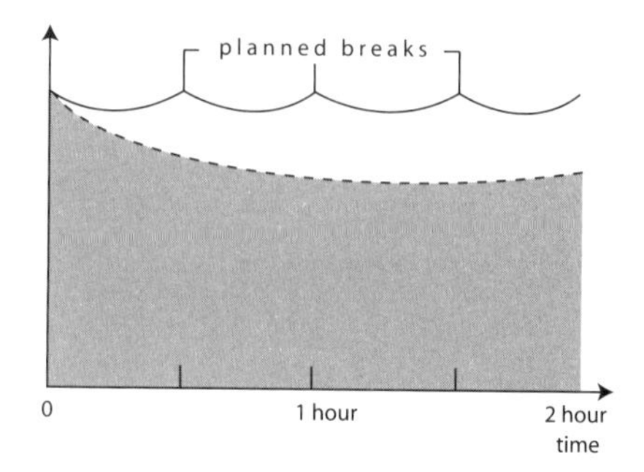

Fig. (ii) Short revision sessions are more effective

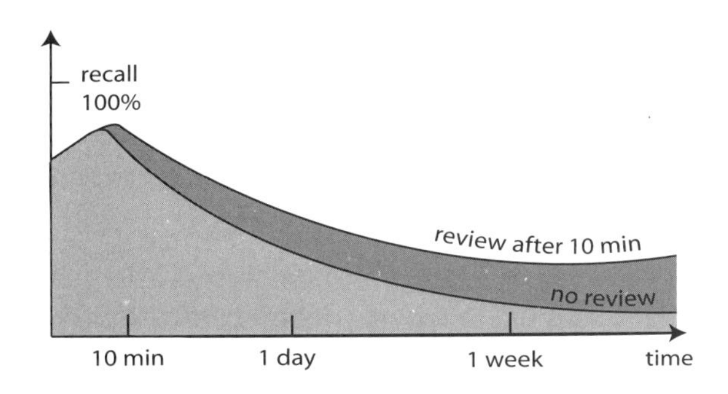

Fig. (iii) Recall falls rapidly after one revision session

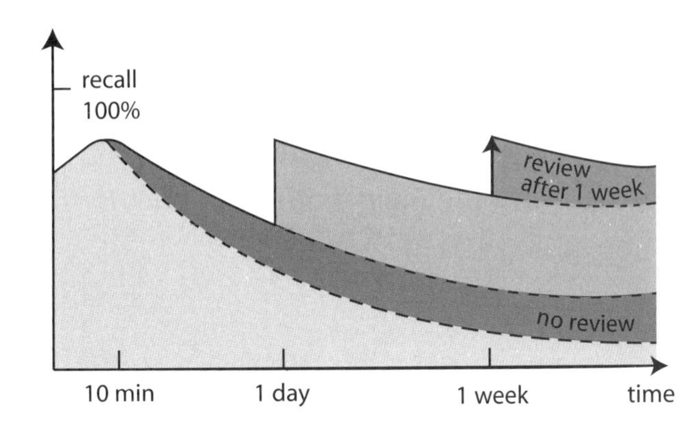

Fig. (iv) Recall improves by reviewing revised material regularly

EXAMS

1. It is important that you are aware of the layout of the paper before the exam so you know which sections or questions you have to answer and the choice you will have within each section.
2. Before the exam, work out how much time you have to spend on each question and stick to this when practicing questions and in the exam.
3. Read each question on the exam paper and underline any important words. Don't rush to do the first question you think you can answer.
4. Make sure to read the whole question, as later parts of a question may be more or less difficult for you.
5. Do not spend too long on any one question or part of a question. If you are stuck leave a space and come back later. **Remember it is more difficult to get 100 per cent on half of the question than to get 50 per cent on all of the required questions**.
6. Be as precise as you can and do not write too much (the exam is testing your biology knowledge not your English skills).
7. Where appropriate use diagrams. These should be neatly drawn with a pencil (do not colour). The diagrams should be of a reasonable size and in proportion, with a title and labels.
8. Use the correct scientific terms.
9. **Make sure your writing is legible.**

There are certain terms which appear frequently in biology questions. It is important that you understand what they mean so that you can answer the questions correctly.

Briefly / Concisely	give the main points in short sentences
Compare	mention similarities and differences
Contrast	mention differences only
Comment	say if the statement is true or false with reasons
Define	give the exact meaning of the term or word
Describe	give a detailed account (often a diagram is useful if not required here)
Discuss	explain and give both sides of the issue
Distinguish	give a definition of both terms showing the differences
Explain	give reasons why something is the case or how something happens
Illustrate	use diagrams
List	give a list of words
Outline	state the main points
State	give the main features in brief form

GRAPHS

When you carry out most experiments you are looking at relationships between two variables. Graphs are frequently used to look at these relationships. If you are given graphs in your exam there are only three things you will be asked to do: **draw it, describe it** or **explain it**.

1. Drawing When you draw a graph remember the following:

(a) Draw all graphs on graph paper using pencil.

(b) Label both axes giving units.

(c) Use a scale which fills most of the graph paper.

2. Describing If you are describing the graph imagine you are talking to a person who cannot see the graph so describe it exactly, giving beginning and ending points.

- Block off the graph (from its start) as the graph line changes (see Fig. (v))
- Make a comment at each change, e.g.
  1. At pH 4 the rate of reaction is low.
  2. As pH changes from 4 to 6 the rate of reaction increases rapidly.
  3. At pH 6 the rate is at its maximum (optimum pH for this reaction).
  4. From pH 6 to 8 the rate of reaction declines quickly.

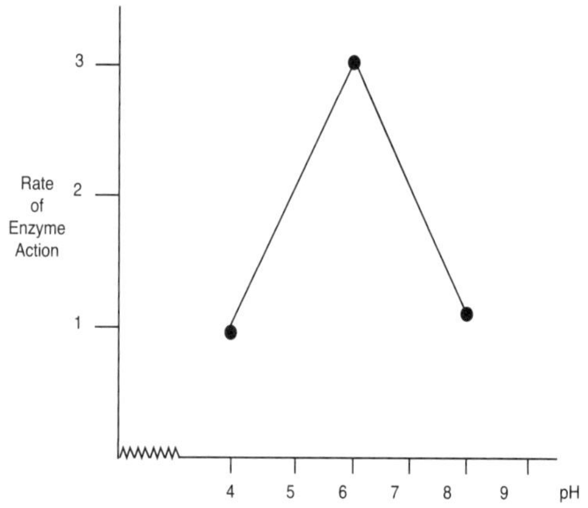

Fig. (v) Graph of ph against rate of enzyme action

3. Explaining When you are explaining a graph use your biology knowledge to explain the relationship the graph demonstrates, i.e.
As the pH changes, the shape of the enzyme changes. At a pH of 6 the enzyme shape allows it to work most effectively. As the pH changes from 6 the shape of the enzyme changes so it loses its function. The further the pH is from 6 the greater the change in shape and hence the greater the loss of function.

PRACTICAL WORK

The Leaving Certificate Biology syllabus includes 22 mandatory practical activities that form an essential part of the course. Through practical work certain important skills are developed.

- You will learn how to assemble and use pieces of equipment (apparatus); for example, how to remove the DNA from cells, how to set up a light microscope and use it to examine some plant cells.
- You will learn the importance of following instructions carefully.
- A biologist learns by observation. By taking readings accurately, by noting the presence and/ or absence of things, by looking for and identifying possible errors, the biologist obtains information about the living world.
- Recording your results, e.g. in a table with the appropriate unit, is very important. Records are proof of what has been done in an investigation.
- Your results and observations need to be interpreted. You might draw a graph of your results and use it to draw a conclusion. A conclusion is a summary of what you have found out. You should be able to explain your conclusion and say how your results agree or disagree with what your investigation set out to do.
- Finally you should evaluate or assess the quality of your investigation by asking questions, such as the following. Were there any errors made? In what ways might errors influence the final results? What precautions could be taken to reduce errors? How did my results compare with those of others in the class?

WRITING UP YOUR PRACTICAL WORK

You should keep a record of your practical work. Try and make your write-ups clear and concise. Be aware of spelling, use correct symbols and units, and ensure your work is legible. A step-by-step approach is a good method of presentation of practical work.

Use the following headings:

1. **Title and date:** state the purpose of the activity, e.g. to show the effect of water on the germination of seeds.

2. **Describe the procedure:** this states what you did and includes the following, where appropriate:
   (a) A clearly labelled diagram of the assembled apparatus.
   (b) Reference to essential adjustments to the apparatus.
   (c) Details of measurements taken during the activity.
   (d) Safety precautions undertaken during the activity.

3. **Data presentation:** all measurements, with the appropriate unit, should be presented in an orderly sequence and set out in a table where appropriate. Reference to a control should be stated when present.

4. Calculations and graphs:

(a) The main steps in any calculations should be set out, so that the process to a final conclusion can be easily followed.

(b) If appropriate, graphs should be drawn (i) to illustrate relationships, and (ii) in the calculation of results.

(c) In drawing graphs (i) every graph should have a title along the top, (ii) suitable scales should be used, (iii) graph axes should be labelled, (iv) points should be plotted and circled clearly, and (v) points should be joined – if the points lie on a straight line then use a ruler; if the points appear to follow a continuous curve then they should be joined using a curved line. Joining up points to form a curve takes practice.

5. Results / observations / conclusions and evaluation: present the results in a clear, scientific manner.

(a) Mathematical results should be expressed in a standard manner with the correct use of units.

(b) Comment on the results obtained.

(c) Refer to possible errors in the activity and state the precautions that can be taken to reduce such errors.

(d) Cross-referencing with other groups should be recorded where appropriate and conclusions drawn.

WORKBOOK FORMAT

The format of the Workbook consists of three sections: A, B and C.

- Section A contains short fill-in style questions relating to Units 1, 2 and 3.
- Section B contains fill-in style questions relating to Mandatory Activities.
- Section C contains longer descriptive-style questions relating to Units 1, 2 and 3.

- Unit 1: Biology – the Study of Life Chapters 1–7
- Unit 2: The Cell Chapters 8–15
- Unit 3: The Organism Chapters 16–33

Note: In each chapter sections B and C appear only where appropiate, i.e. not all chapters contain all three sections.

UNIT 1 chapter 1 *What is Biology?*

SECTION A

1. Complete the following:

(a) Biology is the study of living things.

(b) Zoology is the study of animals.

(c) Ecology is the study of ~~plants and the~~ living things and their environment

(d) The study of inheritance is called genetics.

(e) Give another name for any living thing. organism.

2. List the steps that you can take between making an initial observation and arriving at the formation of a principle or theory in science.

Steps:
Observation, hypothesis, experimentation, collection and interpretation of data, conclusion, relating the conclusion to existing knowledge, reporting and publishing the findings.

What is another name for these steps? scientific method.

3. Explain the following in relation to scientific experiments:

(a) The use of a control.
it is used to provide a standard against which the actual experiment can be judged.

(b) The need to take a large number of samples.
it reduces the risk that the results are due to individual differences, rather than the factor being tested.

(c) The need to repeat experiments.
An experiment must be repeated so as to make sure that the findings are correct and were not just a 'fluke'.

(d) Double-blind testing.
Neither the person ~~not~~ being tested nor the tester should know who is recieving the real treatment or who is recieving the placebo

4. Identify the following safety symbols:

(a)

(b)

(c)

(d)

(e)

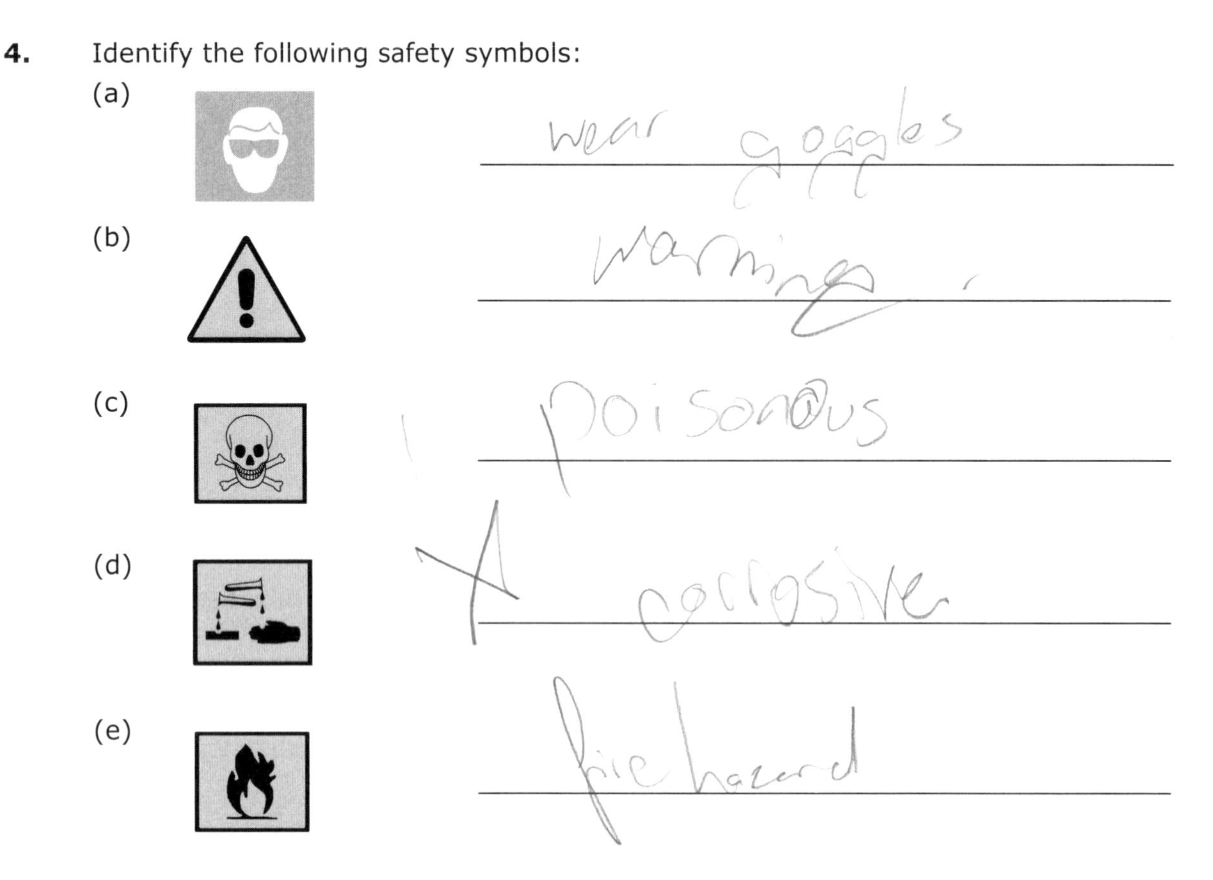

5. Use a simple biology experiment with which you are familiar and explain the difference between the 'control' and the 'experiment'.

6. The discovery of the antibiotic penicillin in 1928 by Sir Alexander Fleming is an example of a 'chance discovery'. Using resource books, CDs such as *Encarta,* and the internet, find out more about Sir Alexander Fleming's famous discovery.

chapter 2 *The Characteristics of Life*

SECTION A

1. Match the words in the left-hand column with those in the right-hand column.

Excretion	Responding to a stimulus
Nutrition	Made up of cells
Metabolism	Getting rid of harmful wastes
~~Behaviour~~ Response	Producing offspring
Organisation	Feeding
Reproduction	Making energy in cells

2. State, in column 2, the characteristic of life that best matches the example listed in column 1. Choose from the following characteristics of life: organisation; nutrition; excretion; ~~behaviour~~ response; reproduction.

Column 1	Column 2
Jumping with fright	response
A dog urinating against a tree	excretion
Laying eggs	reproduction
Having breakfast	nutrition
Using up energy playing a match	organisation.

3. (a) Place the following in the correct order, starting with the smallest:

organ system tissue organism cell

Cell → tissue → organ → System → organism.

(b) Place each word from the following list into the correct column in the table below:

ant bone brain blood ear egg flower frog gut greenfly
heart leaf lung muscle phloem sperm stem xylem

Cell	Organ	Organism	Tissue
Sperm	lung	greenfly	muscle
egg	heart	frog	blood
xylem	gut	flower	bone
	brain	ant	stem
	ear	leaf	bone

4. (a) From where do animals obtain the nutrients they need?

other plants or animals.

(b) From where do plants obtain the nutrients they need?

plants make their own food

(c) What is a food chain?

a food chain is the linkage of animals and plants in the ranking of food

(d) Give an example of a simple food chain. ______

(e) Why do you think a food chain usually starts with a plant?

5. (a) What is meant by the term excretion?

(b) What is meant by the term nitrogenous waste?

Give an example of nitrogenous waste. ______

(c) Name two things that plants excrete.

(d) Name two things that animals excrete.

6. Distinguish between the following pairs of terms by briefly explaining each term and give an example where appropriate.

(a) Unicellular and multicellular organisms.

(b) Tissue and organ.

(c) Autotroph and heterotroph.

(d) Sexual and asexual reproduction.

7. (a) Define the term metabolism. ______

Give examples of metabolism and outline the importance of metabolism for living things.

(b) Distinguish between a 'living', a 'non-living' and a 'dead' thing, giving named examples in each case.

All living things must display the 5 characteristics. If they display some they can be called 'non living', if they display none they are 'dead'.

(c) What do you understand the term 'life' to mean? ______

does

chapter 3 *Introduction to Ecology*

SECTION A

1. Complete the following:

(a) The study of the interactions between groups of organisms and their environment is known as ______

(b) A community of organisms interacting with one another and with their surroundings is called an ______

(c) The place where an organism lives is known as its ______

(d) An organism that eats only plant material is called a ______

(e) An organism that makes its own food is known as an ______

(f) The role of an organism in its habitat describes its ______

2. Distinguish between the following pairs of terms:

(a) Ecology and ecosystem. ______

(b) Herbivore and carnivore. ______

(c) Biotic and abiotic factors. ______

(d) Food chain and food web. ______

(e) A grazing food chain and a detritus food chain. ______

3. Use the information in the food chain below to answer the following questions:

green plants → caterpillars → thrushes → sparrowhawks

(a) Which organism is the producer? ______

(b) Name the primary consumer. ______

(c) Name the secondary consumer. ______

(d) Name the tertiary consumer. ______

(e) Which organisms are carnivores? ______

(f) Draw a pyramid of numbers to represent the food chain.

4. (a) What does a pyramid of numbers show? ______

______

(b) What is the approximate percentage of energy lost at each transfer within the food chain?

______

(c) In what form

(i) does energy enter a grazing food chain? ______

(ii) is energy lost from the food chain? ______

(d) Examine the following pyramids of numbers and match each one to the correct food chain which it represents (Fig. 3.1).

Food chains:

Grass → sheep → fleas ________

Rose bush → greenflies → small birds ________

Microscopic algae → water fleas → pond skaters ________

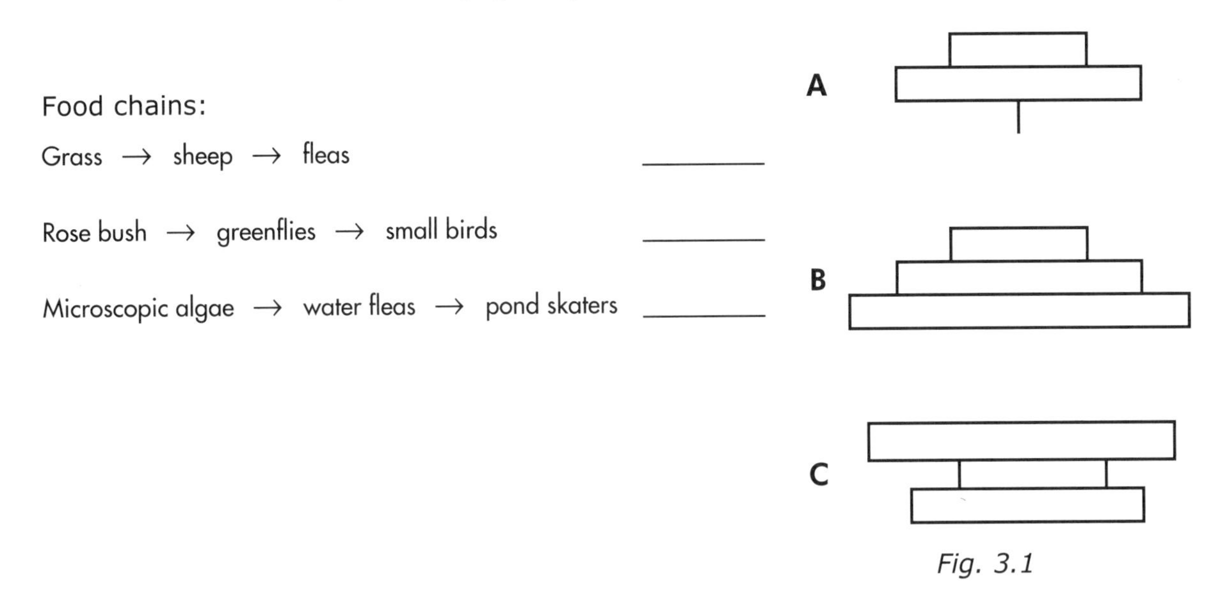

Fig. 3.1

5. The diagram shows a pyramid of numbers from an ecosystem (Fig. 3.2). Use the diagram to answer the following questions.

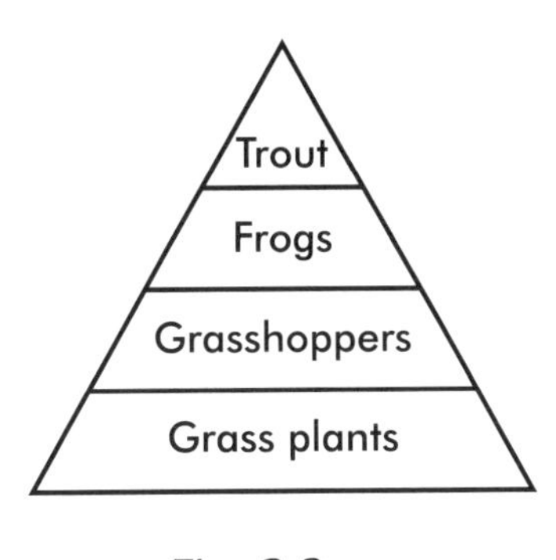

Fig. 3.2

(a) Which organism in this pyramid is an autotroph?

Grass ☐ Grasshoppers ☐
Frogs ☐ Trout ☐

(b) The greatest amount of energy in this pyramid is found in the:

Grass ☐ Grasshoppers ☐
Frogs ☐ Trout ☐

(c) The food chain illustrated in this pyramid is:

trout → frogs → grasshoppers → grass ☐ grasshoppers → grass → frogs → trout ☐
grass → grasshoppers → frogs → trout ☐ frogs → grasshoppers → trout → grass ☐

(d) The primary consumer in this pyramid is:

Grass ☐ Grasshoppers ☐
Frogs ☐ Trout ☐

(e) This pyramid suggests that in order to live and grow, 100 frogs would require:
(i) less than 100 grass plants ☐
(ii) 100 grasshoppers ☐
(iii) more than 100 grasshoppers ☐
(iv) no grasshoppers ☐

6. (a) Give the chemical formula for a molecule of carbon dioxide ____________

(b) What percentage of the air is carbon dioxide? ____________

(c) Name two processes which add carbon dioxide to the atmosphere.
(i) ____________ (ii) ____________

(d) The diagram represents the main components of the carbon cycle in nature (Fig. 3.3). Let the letter P represent photosynthesis; C represent combustion (burning); and R represent respiration. Place the letters P, C and R on the correct arrows on the diagram.

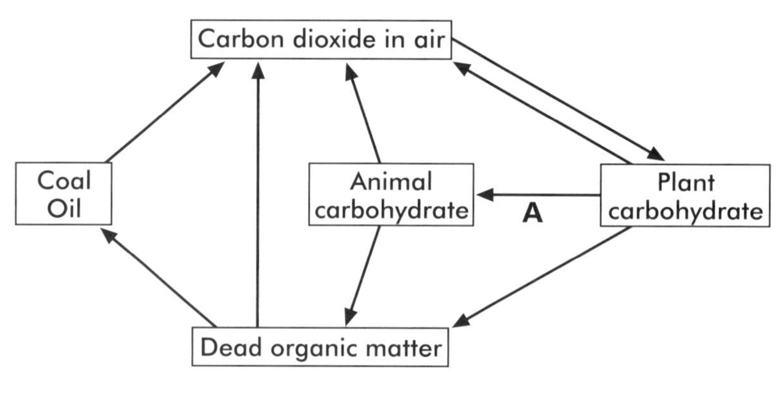

Fig. 3.3

7. The diagram shows part of the nitrogen cycle (Fig. 3.4).

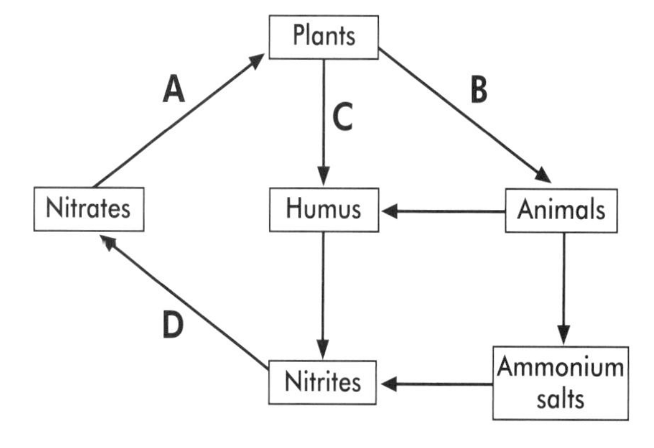

Fig. 3.4

(a) Name a type of organism that can convert ammonium salts to nitrates.

____

(b) The nitrogen cycle involves a number of processes. In the table below write the name of a process occurring at each labelled arrow.

Label	Process
A	
B	
C	
D	

SECTION C

1. The diagram represents part of a food web for the organisms which live in an aquarium (Fig. 3.5).

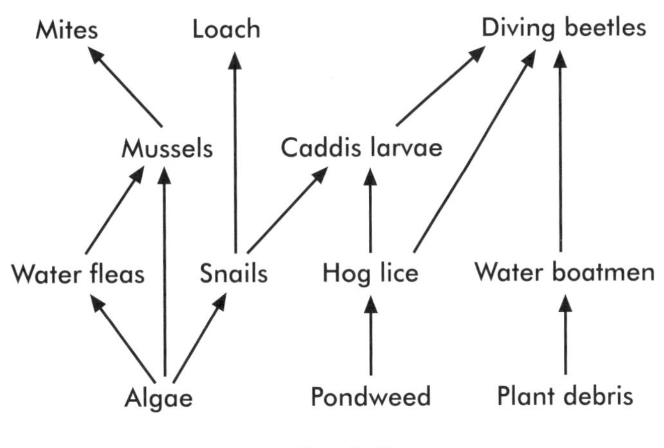

Fig. 3.5

(a) From the food web give (i) a primary producer, (ii) a primary consumer, (iii) all the secondary consumers and (iv) a food chain which contains four organisms.

(b) There are three small fish in the aquarium. These fish, which are about 4 cm in length, feed on snails. There are 35 snails in the aquarium. These are about 5 mm in diameter and they feed on the single-celled microscopic algae. Construct and label a pyramid of numbers for the above organisms.

2. Comment on the validity of the following statements.

(a) Organisms tend to get smaller in size as you go up through a grazing food chain.

(b) The number of organisms decreases at each trophic level.

(c) The territories of top carnivores are usually very large.

(d) There are usually more than five links in a food chain.

3. Read the following passage and answer the questions which follow:

In general no two species of organism can occupy the same niche in the same habitat if they have the same requirements. Organisms can live in the same habitat if they occupy different niches within the habitat. For example, the bank vole and the field mouse share the same habitat but the bank vole is diurnal and the field mouse is nocturnal.

The turnstone and the curlew are birds that both feed on mud flats. This is possible because the turnstone has a short beak and feeds on animals that live on or near the surface, such as crabs and snails, whereas the curlew has a very long curved beak which enables it to reach deep-lying prey, such as lugworms.

(a) Give the meaning of the underlined words.

(b) How is it possible that the curlew and the turnstone are able to live in the same habitat?

(c) Using organisms from the passage and your knowledge of food chains, construct a simple food chain.

4. (a) Why do organisms need nitrogen?

(b) Outline the problems animals have in obtaining the nitrogen they need.

(c) Describe how nitrogen in the air is converted into a form suitable for use by plants.

chapter 4 *Ecological Relationships (Higher Level Only)*

The material covered in this chapter is for higher level only.

SECTION A

1. Define the following terms:

(a) Population ______________________________

____

(b) Competition ______________________________

____

(c) Predation ______________________________

____

(d) Parasitism ______________________________

____

(e) Symbiosis ______________________________

____

2. Distinguish between the following pairs of terms:

(a) Population and community ______________________________

____

(b) Immigration and emigration ______________________________

____

(c) Scramble and contest competition ______________________________

____

(d) Predator and prey ______________________________

____

(e) Endoparasites and ectoparasites ______________________________

____

3. List three things that cause a population to grow and three things that cause a population to stop growing.

Causes of population increase	Causes of population decrease
1.	1.
2.	2.
3.	3.

4. Give a biological explanation for each of the following:

(a) Hyenas live in packs (groups). ______________________________

____

(b) Hover flies have no sting but have yellow stripes on their bodies similar to wasps.

____

(c) Parasites generally do not kill their host. ______________________________

____

(d) Water lilies have stomata on their upper surface and the plants float on water.

____

____

(e) Tapeworms have hooks and suckers on their heads and produce millions of eggs.

____

____

SECTION C

1. The following chart shows the feeding relationships of a series of organisms (Fig. 4.1). State whether each of the following changes is likely to cause an increase or a decrease in the population of O. Explain your answer in each case.

(a) An increase in the population of M.
(b) An increase in the population of N.
(c) A decrease in the population of Q.
(d) An increase in the population of R.
(e) An increase in the population of P.

P Q R
N O
M
L

Fig. 4.1

2. A suspension of yeast cells was added to dilute sucrose solution at 25 °C. Over a 16-day period 10 cm^3 samples were withdrawn each day and the number of organisms counted. On the fourth day a small number of a unicellular organism (*Paramecium*) was added. The results were as shown in the table below.

(a) Plot two graphs, on the same axes, to show the numbers of yeast and *Paramecium* in the samples over the 16-day period. Use graph paper and plot time on the horizontal axis.

(b) Outline the likely cause of the change in the population of *Paramecium* between (i) days 5 and 6, and (ii) days 7 and 8.

(c) Outline the likely cause for the change in population of yeast cells between days 7 and 8.

Days	No. of yeast cells in 10 cm^3 sample	No. of *Paramecium* in 10 cm^3 sample
1	20	
2	86	
3	220	
4	264	
5	266	32
6	224	150
7	114	168
8	156	75
9	220	72
10	216	135
11	124	165
12	85	94
13	180	52
14	177	90
15	120	143
16	56	122

3. Despite fluctuations in some countries the human population continues to grow. At present there is no sign of the world population levelling off.

(a) Suggest two reasons to explain why the human population is increasing.
(b) Explain why the human population fluctuates.
(c) Suggest two ways in which the rate of increase of the human population can be halted.

4. Using TV, books, newspapers and the internet find out:
(a) Two places where there is famine in the world today, and
(b) Two places where there is a war at the moment.

In each case outline the cause(s) and effects of the famine and war.

chapter 5 *The Effects of Humans on the Environment*

SECTION A

1. Define the term pollution.

____

2. Complete the following table:

Type of pollution	Definition	Example
Domestic		
Agricultural		
Industrial		

3. Give **two** examples of **either** water pollution **or** air pollution and state one harmful effect of each example on the environment. State whether the examples are of water pollution or air pollution.

Example 1 ______________________________

Type of pollution ______________________________

Harmful effect ______________________________

Example 2 ______________________________

Type of pollution ______________________________

Harmful effect ______________________________

4. Distinguish between the following pairs of terms:

(a) Pollution and pollutant. ______________________________

____

____

(b) Pesticide and insecticide. ______________________________

____

____

5. The map shows acid rain in Western Europe (Fig. 5.1).

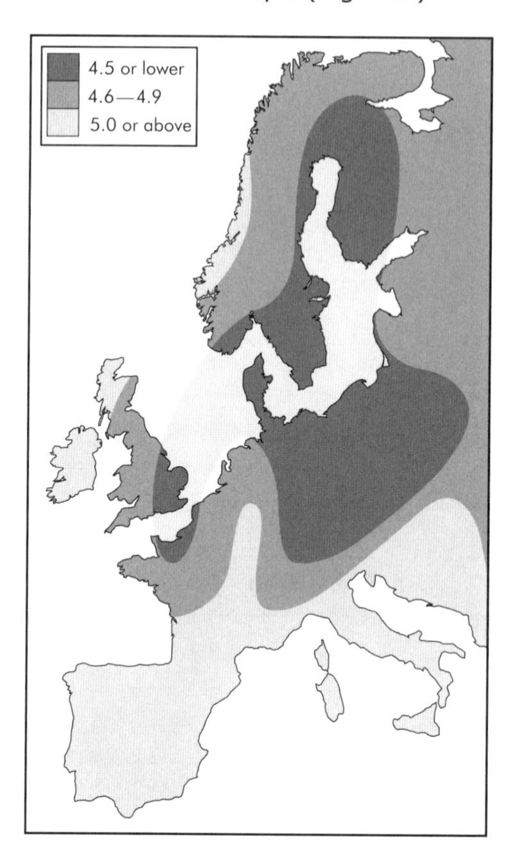

Fig. 5.1

(a) How is acid rain formed?

____

(b) What effect does acid rain have on plants?

____

(c) Explain why the pH of the rain falling on Ireland is different from that falling on Scandinavia.

____

(d) In Sweden more than 4000 lakes no longer contain live fish. Some of the lakes have become 100 times more acidic over the past 30 years. Explain why so many of the lakes are 'dead'.

____

(e) State two ways in which acid rain pollution could be reduced.

(i) ____

(ii) ____

SECTION C

1. (a) Explain what conservation of the environment means.

(b) 'Zoos, botanic gardens and seed banks have an important role to play in conservation.' Do you agree with this statement? Give reasons for your answer.

2. The rubbish we place in our waste bins is made up of biodegradable and non-biodegradable materials.

(a) What does 'biodegradable' mean?

(b) Sort the following waste materials under the headings biodegradable and non-biodegradable:

apple skins egg shells baked bean cans plastic bags tea leaves glass bottles lettuce leaves

(c) Recently a local county council decided to encourage people to compost their household rubbish. It was thought that 25 per cent of household rubbish could be composted. Each householder was given the opportunity to purchase a composter into which they would put all their organic waste.

(i) What is a composter?

(ii) State two ways in which composting household rubbish could be of benefit (1) to the householder and (2) to the environment.

(d) Name one type of organism which is involved in the process of decay.

3. Explain the following:

(a) Waste management.

(b) Landfill sites.

(c) Bring banks.

(d) Waste minimisation.

(e) Incineration of waste.

chapter 6 *The Study of an Ecosystem*

SECTION A

1. Complete the table below for five animals found in your named habitat.

Name of habitat: ______________________________

Name of animal	Classification group	What it eats	What eats it	Adaptation to life in the habitat

2. Complete the table below for five plants found in your named habitat.

Name of habitat: ______________________________

Name of plant	Classification group	Colour (of flower if present)	What eats it	Adaptation to life in the habitat

SECTION B

1. Below is a key used to identify some butterflies and moths.

1.	Antennae feathery	Go to 2
	Antennae clubbed (not feathery)	Go to 6
2.	Both pairs of wings similar in size	Go to 3
	Front pair of wings much larger than rear wings	Go to 4
3.	Both pairs of wings patterned	Go to 5
	Only front wings patterned	Alder Moth
4.	Body 2.5–3.0 cm long	Elephant Hawk Moth
	Body 3.0–3.5 cm long	Privet Hawk Moth
5.	Wings with large eye spots	Emperor Moth
	Wings without eye spots	Gipsy Moth
6.	Dark wings with light-coloured edges	Camberwell Beauty Butterfly
	Light wings with dark markings	Small White Butterfly

(a) Use the key to identify the insect in the drawing (Fig. 6.1).

____

(b) Using the information in the key, state one difference between butterflies and moths.

____

____

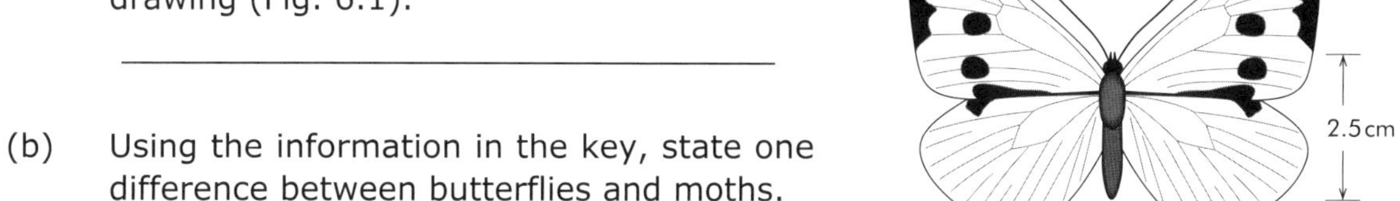

Fig. 6.1

(c) The Emperor Moth has wings with large eye spots. Using information in the key, state two other features of the Emperor Moth.

____

____

2. The drawings show a selection of Arthropods living in a wood.

1.	With wings	Go to 2
	Without wings	Go to 3
2.	First part of abdomen thin, forming a waist	*Apanteles*
	No waist between thorax and abdomen	*Tenthredo*
3.	Ten legs or less	Go to 4
	More than 10 legs	*Uthobius*
4.	Six legs	*Agriotes*
	Eight legs	*Pergamasus*

(a) Use the key to identify the Arthropods A–E (Fig. 6.2). Write the letter of each Arthropod next to its name below.

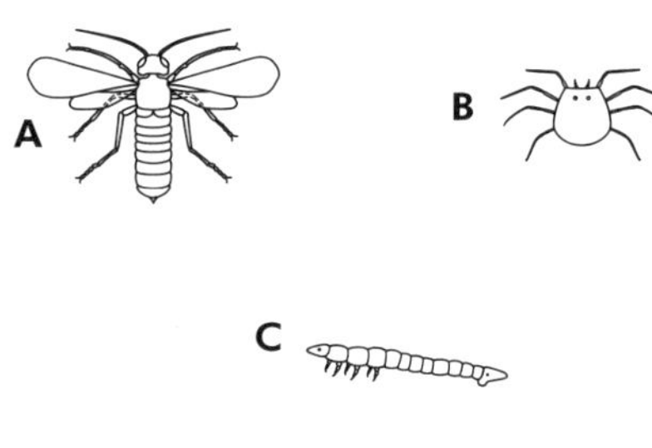

Agriotes ______________________

Apanteles ______________________

Lithobius ______________________

Pergamasus ______________________

Tenthredo ______________________

Fig. 6.2

3. Name each piece of equipment shown in the diagram opposite (Fig. 6.3). Briefly describe how each piece of equipment is used and name an organism that can be collected using each one.

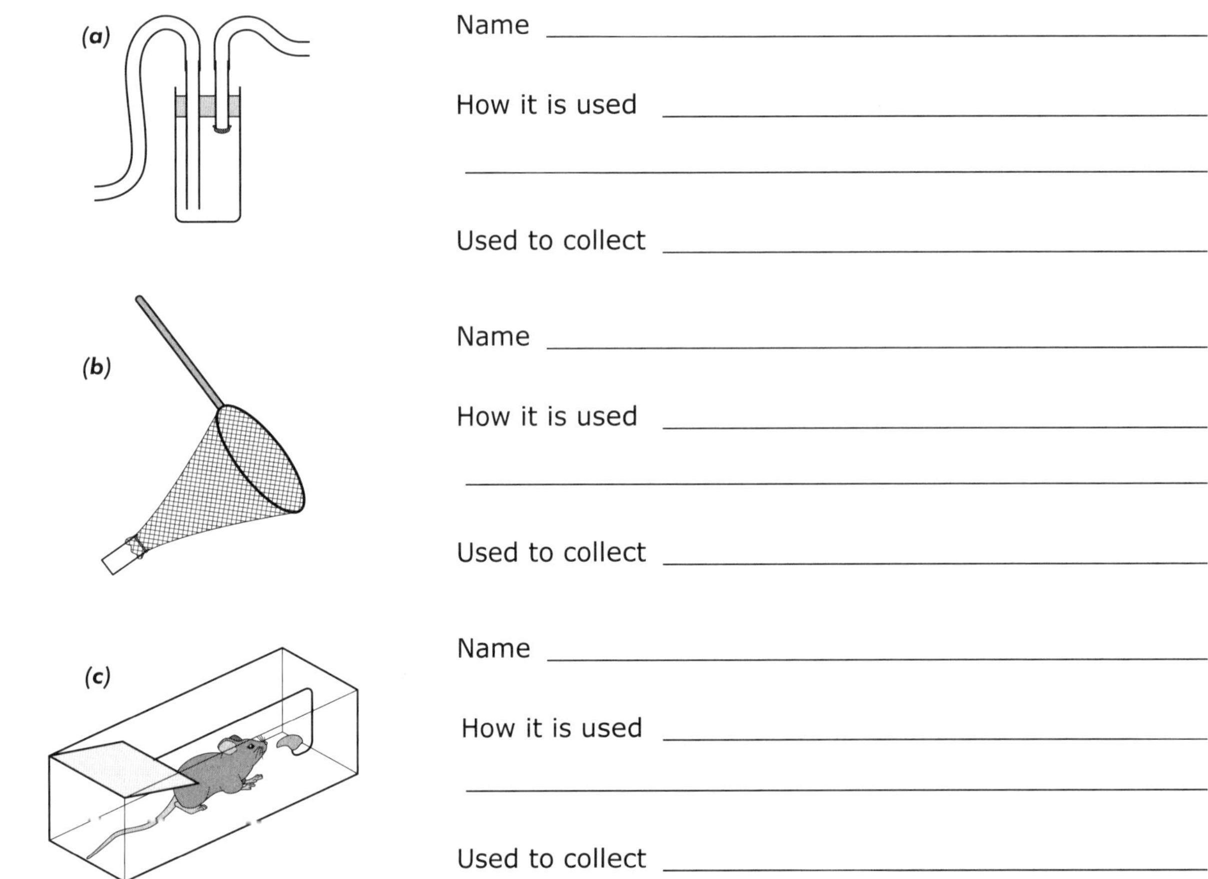

(a)

Name ______________________

How it is used ______________________

____

Used to collect ______________________

(b)

Name ______________________

How it is used ______________________

____

Used to collect ______________________

(c)

Name ______________________

How it is used ______________________

____

Used to collect ______________________

Fig. 6.3

4. (a) In the space provided draw a labelled diagram of a Tullgren funnel.

(b) Describe what a Tullgren funnel is used for. ______________________

____

(c) Outline how the funnel works. ______________________

____

SECTION C

1. Ten important features of an organism's physical environment are: (1) light, (2) temperature, (3) air currents, (4) rainfall, (5) pH, (6) salinity, (7) light penetration of water, (8) water flow, (9) percentage (%) dissolved oxygen and (10) humidity.

(a) Choose and name a habitat you have studied and explain how you would measure any five of the features on the list.

(b) Which of the features do you consider to be the most important in your habitat? Give a reason for your answer.

(c) Name one organism which lives in the habitat and explain how its life is affected by each feature on the list.

2. A total of 50 periwinkle snails were collected at random from an area of rocks on the seashore. Each periwinkle was marked on the shell with a spot of enamel paint. The periwinkles were then returned to the rocks. Five days later a second random collection of 40 periwinkles was made in the same part of the shore. Of the 40 collected, four were found to be marked.

(a) Calculate the population of periwinkles in the area.

(b) Describe, giving reasons, the way the animals should be marked.

(c) What name is given to this method of estimating the population of animals?

3. An area of hillside 140 m long and 60 m wide was chosen for study. A dirt track 3 m wide cut through the full length of the area. In order to determine the number of ferns growing on the hillside 20 quadrats, each of edge 0.5 m, were thrown at random. The table below shows the numbers of fern in each quadrat. No ferns were growing on the dirt track, and no quadrats were thrown there.

Quadrat number	1	2	3	4	5	6	7	8	9	10	11	12	13	14	15	16	17	18	19	20
Number of fern plants present	3	2	0	0	5	2	0	1	3	4	4	2	0	0	2	5	3	2	0	2

(a) Calculate the percentage frequency of ferns on the hillside.

(b) Calculate the total number of ferns growing on the hillside.

(c) What is the reason for 20 quadrats being thrown and not 2 or 200?

4. A point quadrat was used to determine the percentage cover of clover plants in a meadow. The point quadrat has 10 holes and it is used six times. Clover is recorded 24 times. What is the percentage cover of clover plants?

chapter 7 *The Chemicals of Life*

SECTION A

1. Give the chemical symbol for each of the following elements:

Name	Symbol
Hydrogen	
Calcium	
Nitrogen	
Potassium	
Iron	
Carbon	
Oxygen	
Sodium	
Magnesium	
Iodine	

2. Give the name of the chemicals whose symbols are:

Symbol	Name
C	
K	
Cl	
P	
Na	
Zn	
Ca	
O	
S	
Fe	

3. (a) What is a biomolecule? ______________________________

____

(b) Give two main functions of food in our body.

(i) ______________________________

(ii) ______________________________

(c) List the six components of a balanced diet. Carbohydrates, Proteins,

Fats, Vitamins, Minerals, Water

(d) State two functions of water in the body.

(i) ______________________________

(ii) ______________________________

(e) Why is respiration described as a catabolic reaction? ______________________________

____

4. Complete the following table about carbohydrates, fats and proteins.

Food type/feature	**Carbohydrates – (mono and disaccharides)**	**Carbohydrates – polysaccharides**	**Fats**	**Proteins**
Elements present			C, H and O	
Sub-units involved				
Solubility in water				Some, not all
Named example	Glucose			
Function in the diet of named example				
Good food source			Butter	
Laboratory test(s)	Benedict's test			

5. (a) Name the element present in proteins but not in lipids. ____________________

(b) The diagram shows a lipid molecule and a protein molecule (Fig. 7.1).

1.

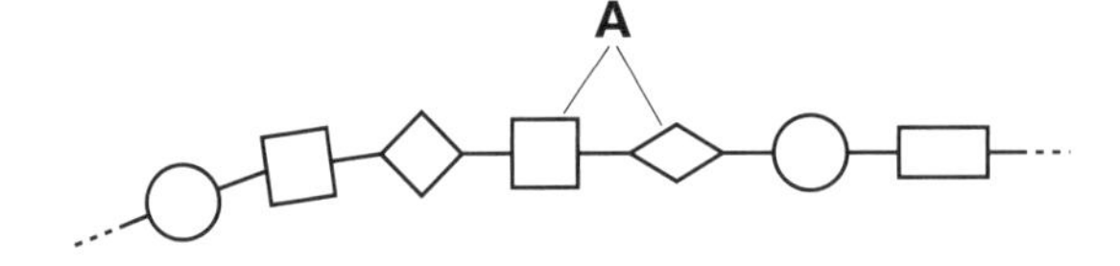

2.

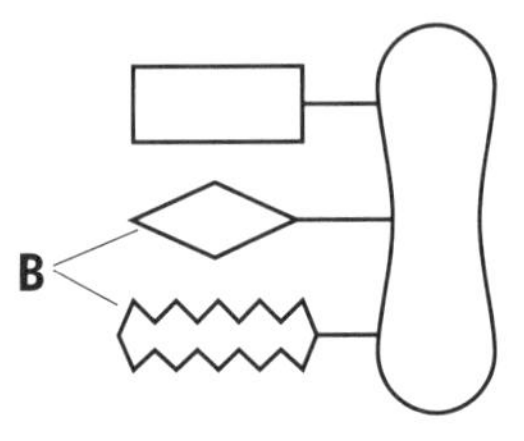

Fig. 7.1

(i) Which diagram 1 or 2 represents a lipid molecule? ____________________

(ii) Name the structural units A and B.

A ________________________________

B ________________________________

(c) In the space provided sketch a similar model of a carbohydrate molecule and label the structural unit in it.

6. From the list below select an example to illustrate each of the following:

vitamin D protein cellulose glucose starch
vitamin C calcium iron magnesium sucrose fat

(a) A structural carbohydrate ______________________

(b) An essential component of haemoglobin ______________________

(c) A vitamin manufactured in human skin ______________________

(d) A mineral involved in tooth structure ______________________

(e) A reducing sugar ______________________

(f) An important component of cell membranes ______________________

(g) A water-soluble vitamin ______________________

(h) Essential for growth and repair in the body ______________________

7. Match the biomolecules in the left-hand column with those in the right-hand column.

Biomolecule	**Function**
Vitamin C	Acts as a store of carbohydrate in plants
Calcium	Needed for energy in the body
Starch	Necessary to produce connective tissue
Protein	Acts as a store of carbohydrate in animals
Fat	Hardens bones and teeth
Glycogen	Forms hormones

8. Answer the following:

(a) All proteins contain the elements carbon, hydrogen, oxygen and______________________

(b) Benedict's solution is used to test for ______________________

(c) Scurvy is caused by a lack of ______________________

(d) Fat is needed in the diet for ______________________

(e) The storage carbohydrate in plants is called ______________________

(f) The function of roughage in the diet is to ______________________

(g) List three foods that are rich in protein ______________________

(h) Fats consist of smaller units called ______________________

SECTION B

1. Complete the following table.

	Reducing sugar	**Starch**	**Fat**	**Protein**
Test reagent(s)				
Is heat needed?				
What is the colour change, if result is positive?				

2. In the space provided below, draw a labelled diagram of the apparatus you could set up to investigate the presence of a reducing sugar in a sample of food. Describe the steps you carried out in the activity.

3. A student was given five solutions, A, B, C, D and E. Each solution was tested for the presence of the following:

Glucose – using the Benedict's test.

Starch – using the Iodine test.

Protein – using the Biuret test.

The final colour observed at the end of each of the tests is shown in the table below.

	Solution A	**Solution B**	**Solution C**	**Solution D**	**Solution E**
Benedict's test	Blue	Red/orange	Blue	Blue	Red/orange
Biuret test	Blue	Blue	Blue	Purple	Purple
Iodine test	Blue-black	Blue-black	Yellow/brown	Yellow/brown	Blue-black

Which solution contained the following?

(a) Protein only. ______________________

(b) Glucose only. ______________________

(c) Starch only. ______________________

(d) All three. ______________________

(e) None of the three. ______________________

(f) Starch and glucose only. ______________________

SECTION C

1. A packed school lunch contains a ham and cheese sandwich made with brown bread and butter, a yoghurt and a quarter-litre carton of blackcurrant juice.

(a) Which of the above foods would be a **good** source of the following?

(i) Carbohydrate.

(ii) Fat.

(iii) Protein.

(iv) Roughage (fibre).

(b) Name one vitamin and one mineral found in this lunch. State which of the above foods contains the vitamin and mineral that you name (you may use more than one food).

(c) Which of the above foods supplies materials used by the body?

(i) To repair an injury.

(ii) To provide the energy to run 100 m.

2. The table below shows the composition and energy content per 100 g of various foods.

Food	Energy (kJ)	Carbohydrate (g)	Fat (g)	Protein (g)
Milk	272	4.8	4	3.3
Chicken	771	0	7.3	29.6
White bread	1060	54	1.7	8.3
Apple	193	12	0	0.3
Sausages	1520	9.5	32	10.6
Fizzy drink (can)	550	30	0	0

Food	Fibre (g)	Vitamin C (mg)	Iron (mg)	Calcium (mg)	Vitamin D (mg)
Milk	0	1	0.1	120	0.1
Chicken	0	0	2.6	15	0
White bread	2.8	0	1.8	100	0
Apple	20	5	0.3	4	0
Sausages	9.5	0	1.1	15	0
Fizzy drink (can)	0	0	0	0	0

Answer the following:

(a) Which food contains (i) the most carbohydrate, and (ii) the least protein? What do you need proteins for?

(b) If the foods were eaten in equal amounts state which food would be most useful in (i) preventing scurvy, (ii) growth, and (iii) caring for teeth and bones.

(c) What do you need fibre for? Which foods give (i) the most fibre, and (ii) the least fibre?

(d) How much energy is there in (i) 100 g white bread, (ii) 50 g of sausages, and (iii) two cans of fizzy drink?

(e) Why do you need iron? Which food has the most iron?

3. A 20-year-old young woman kept a record of her food and drink intake for a single day. The quantities of some major nutrients in the food were calculated, and the totals compared with the average daily requirement of a woman that age. The figures are shown in the table.

Meal	Item	Amount	Energy (kJ)	Protein (g)	Lipid (g)	Carbohydrate (g)	Ca (mg)	Fe (mg)	Vit C (g)
Breakfast	Toast	70 g	900	7	2	50	80	1	0
	Butter	15 g	450	0	12	0	2	0	0
	Jam	30 g	300	0	0	20	4	0	1
	Tea with milk	2 cups	200	2	4	6	100	0	0
Lunch	Sausages	75 g	1150	9	24	10	30	0.5	0
	Chips	300 g	3600	12	30	105	38	3	30
	Fizzy drink	1 can	550	0	0	30	0	0	0
Dinner	Shepherd's pie	120	2800	5	30	8	23	2	2
	Baked beans	220	600	10	1	20	120	3	4
	Apple pie	150	2000	5	25	60	50	1	1
	Ice cream	100	600	4	15	20	125	0	1
Snacks	Crisps	2 pkts	1100	3.2	1720	25	0.8	1	1
	Chocolate	50	1200	5		25	120	1	0
Total per day			15450	62.2	180	379	692.8	12.5	38
Average daily requirement			9400	58	*	*	600	14	30

* Amounts variable

Using the information in the table answer the following questions:

(a) What is the average daily energy requirement of a 20-year-old woman? By how much did the energy content of the day's food intake exceed the average daily energy requirement?

(b) What would happen to the young woman if she continued to exceed the daily energy requirement over a long time?

(c) It is possible that this woman might have an energy requirement much greater than the average. Suggest one reason why this should be so.

(d) Excluding drinks, which of the items eaten during the day had the highest level of calcium per gram of food? State the calcium content of this food in mg/g of food.

(e) Find one nutrient in which the day's food intake is deficient and name a condition that may result if the woman's diet continues to provide too little of the nutrient.

(f) Suggest a menu for an evening meal that would be healthier for the young woman to eat.

4. An experiment was carried out concerning rats and their diet by Sir F.G. Hopkins in the early 1900s. Two groups of young rats, A and B, were used. Both groups were given a diet of carbohydrate, protein, fat, salts and water. In addition, group A were given 3 cm^3 of milk each day. After 18 days, Hopkins switched the addition of milk from group A to group B. The results are shown in the table below.

Time in days	Average mass of rats in group A (g)	Average mass of rats in group B (g)
0	45	45
5	55	48
10	65	51
15	71	50
20	82	45
25	87	50
30	88	62
35	89	67
40	89	72
45	84	78
50	77	84

(a) Plot these results on a single graph, using graph paper. Put time on the horizontal axis.

(b) Mark where the diets changed, by drawing a vertical line up from day 18.

(c) On which days were the masses of groups A and B equal?

(d) Why did Hopkins use two groups of rats rather than just two individual rats?

(e) What effect did the addition of milk have on the growth of the rats in group A during the first 18 days?

(f) After the change in diet on day 18, what happened to the mass of group A and why?

(g) Other than milk, why was it important that the two groups of rats should be given the same amounts of food and water?

(h) One of the conclusions that Hopkins came to was that the milk was providing vitamins. What evidence is there to support this idea?

UNIT 2 chapter 8 *Cell Structure*

SECTION A

1. All living things are composed of ______________________ and all ______________________ were made by pre-existing ______________________

2. The following diagrams show two different cells as seen through a light microscope.

(a) Name the parts labelled:

A ______________________

B ______________________

C ______________________

D ______________________

(b) Identify each of these cells.

X ______________________

Y ______________________

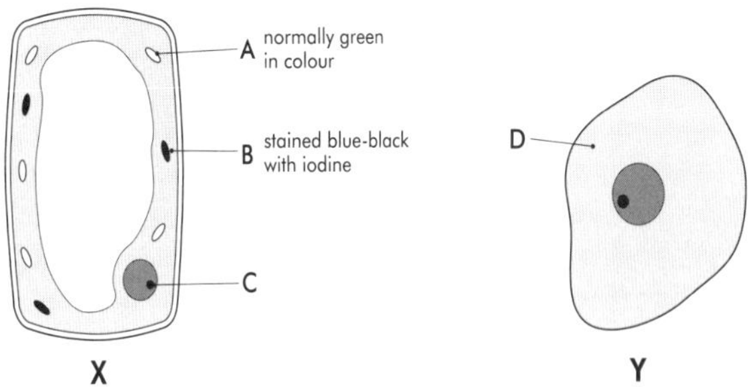

Fig. 8.1

These cells could also be viewed under an electron microscope. What would be the benefit to a scientist in studying these cells using this apparatus?

____

3. (a) Name the structures in the following diagram of an animal cell as seen under an electron microscope.

A ______________________

B ______________________

C ______________________

D ______________________

E ______________________

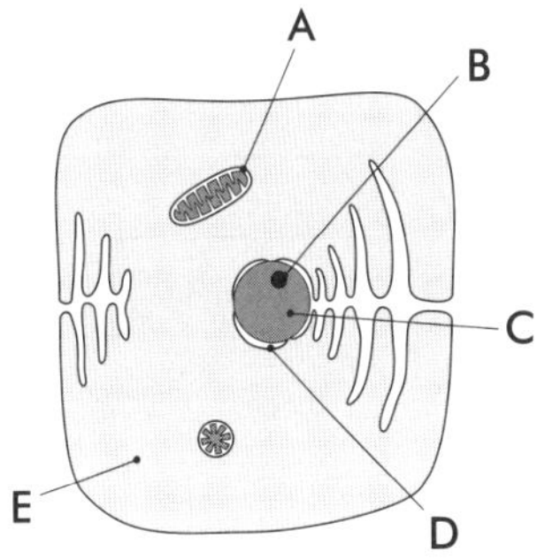

Fig. 8.2

(b) Give a function for each of the parts you have labelled.

A ______________________

B ______________________

C ______________________

D ______________________

E ______________________

(c) What, if any, differences would be seen in a diagram of a plant cell?

____

____

4. Identify the following structures found in cells (Fig. 8.3 a and b).

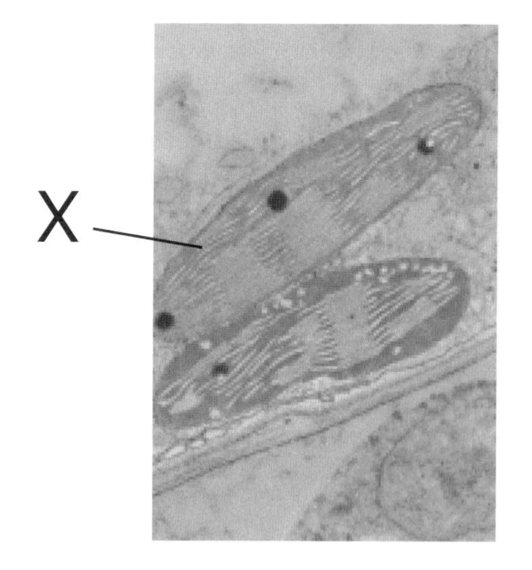

Fig 8.3(a)

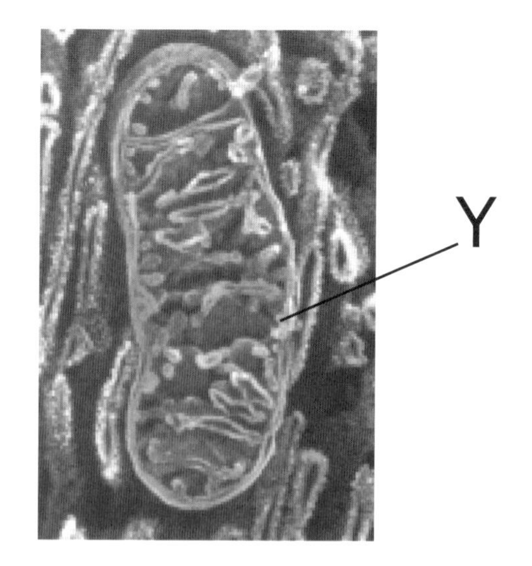

Fig 8.3(b)

X ______________________________

Y ______________________________

What is the function of X? ____

What is the function of Y? ____

5. The following diagram is of a plant cell (Fig. 8.4).

(a) Label the parts:

A ______________________________

B ______________________________

C ______________________________

D ______________________________

E ______________________________

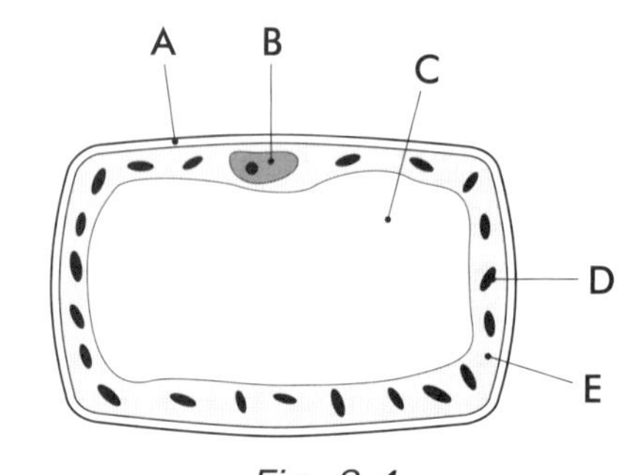

Fig. 8.4

(b) Underline the parts which could also be present in an animal cell.

6. The nucleus contains genetic material called ______________________. This is combined with protein to form ________________. Surrounding the nucleus is a ____________, which contains ______________. These ____________ control the passage of ______________ into and out of the nucleus. Messages made of ______________ leave the nucleus and travel to the ________________ where they can be converted into ________________. All cells in a multicellular organism will have the same _________________ in the nucleus of each cell.

7. (a) List three differences between prokaryote cells and eukaryote cells.

1. ______________________________
2. ______________________________
3. ______________________________

(b) What is the largest group of prokaryotic organisms?

____

SECTION B

1. When comparing a section of onion epidermis for examination under a light microscope, the following steps are taken in sequence. Explain the purpose of each step.

(a) Cut (or pull off) a very thin section.

____

(b) Mount the section in water.

____

(c) Cover the section with a coverslip.

____

(d) Add iodine solution. ______________________________

(e) View the section under low power objective first.

____

2. Label the following parts of a light microscope (Fig. 8.5):

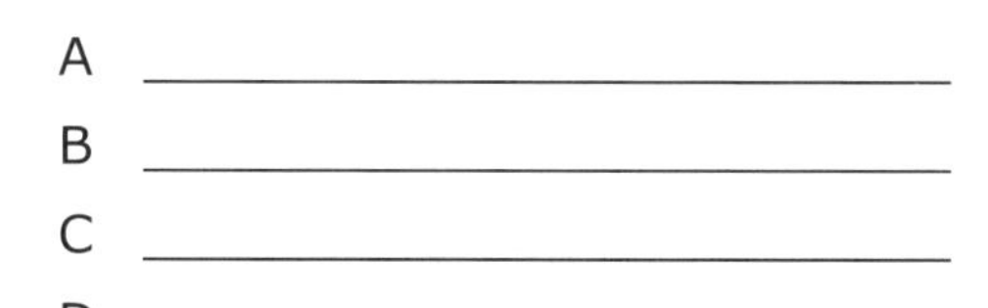

A ______________________________
B ______________________________
C ______________________________
D ______________________________

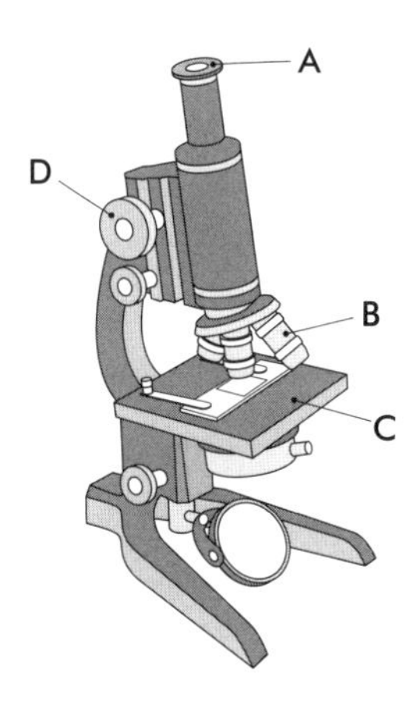

Fig. 8.5

3. (a) Describe how you would prepare an animal cell for examination under a light microscope.

(b) Explain how you would use a light microscope to view your cell.

(c) It is recommended not to use a coarse focus wheel when examining a specimen under high power. Why?

(d) Draw a simple labelled diagram in the space provided of what you would see.

SECTION C

1. (a) Draw a well-labelled diagram of a plant cell as seen under an electron microscope.
(b) What are the differences that would be found in an animal cell?

2. Describe the function and structure of the following cell organelles.

(a) Nucleus.
(b) Chloroplast.
(c) Mitochondria.
(d) Cell wall.

chapter 9 *Cell Metabolism*

SECTION A

1. Fill in the blank spaces from the following list (words may be used more than once):
plants energy sun anabolic respiration glucose metabolism photosynthesis catabolic

Cell ________________ is the term given to chemical reactions that take place in cells. Reactions that produce complex chemicals from simple ones are called _____________ reactions. These reactions usually require _______________. Reactions that breakdown complex chemicals into simpler ones are called _____________ reactions and they usually produce _____________. The energy that is used in all living organisms on this planet comes ultimately from the _____________. The process by which this energy is fixed is called _____________ and it is carried out by all green ____________. This process traps the energy in a chemical called ______________. The energy can be released by cells in a process called ______________.

2. (a) What are enzymes?

____

(b) Enzymes are made of what type of chemical?________________________________

(c) Describe a laboratory test for this chemical.

____

____

____

(d) What is the function of enzymes in living cells?

____

3. (a) What is understood by the term 'activation energy'?

____

(b) What effect do enzymes have on the activation energy?

____

(c) On the diagram below (Fig. 9.1), which illustrates the energy released in a chemical reaction, mark (i) the activation energy, (ii) the energy yield of the reaction, (iii) the substrate molecule and (iv) the products.

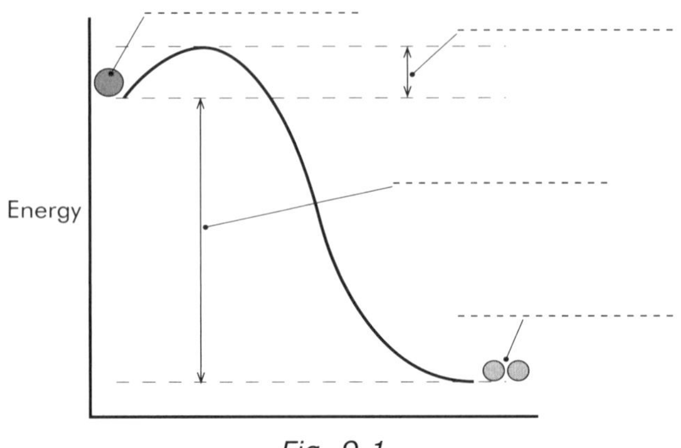

Fig. 9.1

4. The graph (Fig. 9.2) below shows the effect of pH on the rate of reaction of an enzyme.

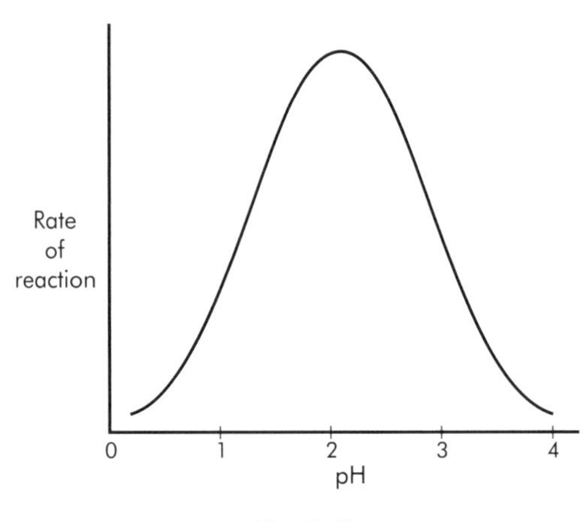

Fig 9.2

(a) Where might you find such an enzyme in a human?

____

(b) What is happening to the enzyme as the pH changes?

____

____

5. In the space below draw a simple graph (Fig. 9.3) to show how the rate of most human enzyme-controlled reactions change with a change of temperature from 0ºC to 40ºC.

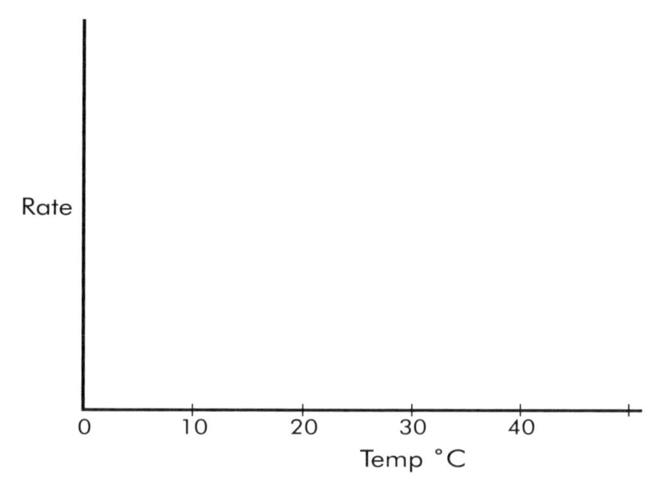

Fig. 9.3

6. (a) What is the advantage of using enzymes in industry?

(b) Why would enzymes be put into gel beads?

(c) What chemical is used to make the gel?

(d) What are the benefits of using gel beads?

7. (a) What is the active site of an enzyme?

(b) What happens to the active site when the enzyme reacts with its substrate?

(c) When the substrate and the enzyme react what is temporarily produced?

(d) At the end of the reaction what is produced?

8. (a) How does a change in pH affect the activity of an enzyme?

(b) What is an 'optimum' pH?

(c) Why does the enzyme function this way at the optimum pH?

(d) What are the optimum pHs of the three enzymes X, Y and Z in the graph below (Fig. 9.4)?

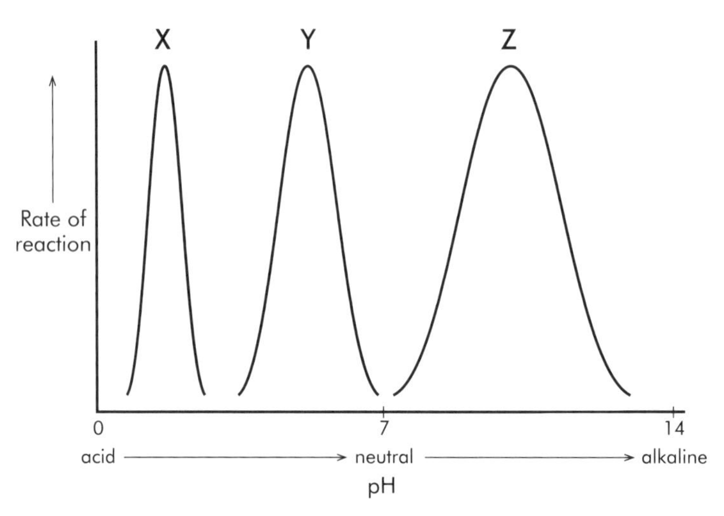

Fig. 9.4

X ________ Y ________ Z ________

9. (a) Why does increased temperature speed up a chemical reaction?

(b) What happens to enzymes if the temperature is raised too high?

(c) What is a denatured enzyme?

(d) How do high temperatures affect cells?

10. (a) What is ATP?

(b) Why do cells use ATP?

(c) What is produced when ATP is broken down? ________

(d) How can ATP be reformed?

___

(e) What two cellular processes produce ATP?

___

11. (a) What is produced in catabolic reactions? ___

(b) What is the function of NAD in these reactions?

___

(c) What is a co-enzyme?___

___

(d) What vitamin is used to make NAD?___

(e) What is the role of NADPH in photosynthesis?

___

___

SECTION B

1. In an experiment four water baths were set up at 10°C, 20°C, 30°C and 40°C, respectively. Into each were placed two test tubes, one containing starch solution and the other containing amylase solution and buffer. After 10 minutes the solutions were mixed and every minute thereafter a drop of the solution was removed from each mixture and tested for the presence of starch.

(a) What is being examined in this experiment?

___

(b) Why were the two test tubes placed into each water bath for 10 minutes before mixing?

___

(c) What is the purpose of the buffer solution?

___

(d) How could you test for the presence of starch?

___

(e) What change would you be looking for in this experiment?

___

___

(f) What results would you expect in this experiment?

____

____

2. In the space provided below draw a labelled diagram of the apparatus you used to investigate the effect of pH on the rate of activity of an enzyme.

Describe the steps you carried out in this experiment, explaining the reason for each step.

____

____

____

____

____

____

____

____

____

____

____

____

____

____

3. In an experiment a solution of sodium alginate was mixed with amylase. The mixture was placed in a syringe and put into calcium chloride drop-by-drop. The resultant beads were placed in a funnel with a tap. A starch solution was dripped onto the gel beads and after 10 minutes the tap was opened and some liquid removed and tested using Benedict's solution.

(a) Give the function of each of the underlined words.

____

____

____

____

____

(b) Describe how you would carry out the test using Benedict's solution.

____

____

(c) What result would you expect? ______________________

(d) Why might this process be used in industry?

____

____

(e) If this apparatus was to be used to show fermentation what modifications would you make?

____

____

____

4. (a) What is meant by the term 'denaturation'?

____

(b) In an experiment to demonstrate denaturation, a test tube containing amylase was boiled for 10 minutes. Why?

____

____

(c) What solution was added to the amylase? ______________________

(d) Why was a buffer solution added to the mixture?

____

____

(e) What control was used in this experiment?

____

____

(f) How did this act as the control?

____

____

(g) What tests were carried out on the two solutions at the end of the experiment?

____

(h) Describe how you would carry out these tests.

____

____

____

____

(i) In the space provided draw a diagram of the apparatus used in this experiment.

SECTION C

1. In an experiment the effect of pH on the action of salivary amylase on starch was examined. The following results were produced.

pH	4	5	6	7	8	9	10
Time (min)	>15	10	7	3	6	9	>15

(a) Draw a graph of the results plotting rate (1/time) on the Y-axis and pH on the X-axis.

(b) At what pH does this amylase work best?

(c) How would you carry out this experiment?

2. Describe how you would carry out an experiment to demonstrate the immobilisation of an enzyme or *Saccharomyces cerevisae*. What are the benefits of this procedure to industry?

3. Describe the action, and explain the importance, of enzymes in living cells. How will temperature and pH affect the action of enzymes?

4. (a) Describe the action of enzymes using the 'induced fit' theory. How does this theory explain the effect of pH on enzymes and what has happened to an enzyme that is denatured?

(b) Describe an experiment to demonstrate heat denaturation of an enzyme and indicate the results you would expect to get.

5. (a) Explain the important roles that ATP, NAD and NADP have in cells.

(b) Biological washing powders have enzymes that come from bacteria. These enzymes are extremely heat tolerant and can be used in temperatures of 60ºC. These washing powders are particularly good at removing food stains and other organic stains, like grass, from clothes.

(i) What function do you think these enzymes have in the bacteria?

(ii) What would happen to many enzymes at 60ºC.

(iii) Can you suggest any other factor to which these enzymes might be tolerant?

(iv) Given agar containing milk or starch and a solution of a biological washing powder, design an experiment to demonstrate the digestive action of these enzymes.

chapter 10 *Biochemical Reactions*

SECTION A

1. (a) Photosynthesis is a process in which sunlight is used to____________________

(b) What are the two reactants of photosynthesis? ____________________

____

(c) What are the two products of photosynthesis?____________________

____

(d) What must be present in the plant for photosynthesis to occur? ______________

____

(e) Give a simple balanced equation for photosynthesis.

____

2. (a) Label the following diagram of a cross section through a leaf (Fig. 10.1).

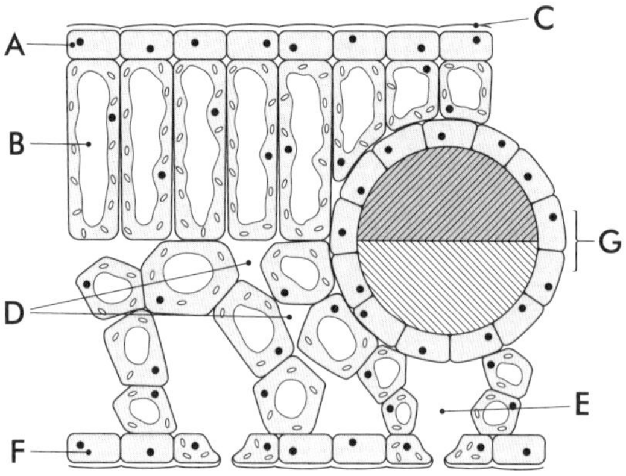

A ____________________

B ____________________

C ____________________

D ____________________

E ____________________

F ____________________

G ____________________

Fig. 10.1

(b) (i) In which cells do you find the largest number of chloroplasts?

____

(ii) Why is this the case?____________________

(c) What is the purpose of the holes in the lower part of the leaf?

____

(d) What is the role of the veins in the leaf?

____

3. (a) What happens in the process of respiration?

____

____

(b) What are the products of aerobic respiration?____________________

(c) Write a simple balanced equation for aerobic respiration.

____

(d) Where in the cell do the two stages of respiration occur?

(i) ____

(ii) ____

(e) What are the differences between aerobic respiration and anaerobic respiration?

____

____

(f) What are the problems associated with the waste products of anaerobic respiration?

____

____

(g) Write a simple balanced equation for anaerobic respiration.

____

4. (a) What is fermentation? ____________________

(b) Where in industry is this reaction used by humans?____________________

(c) What are the benefits of pasteurising milk? ____________________

____

____

(d) How is this process carried out? ____________________

____

____

(e) In the manufacture of yoghurt, milk is pasteurised and then cooled to 46 ^{o}C. Why is it cooled to this temperature?

____

(f) What is contained in the starter culture that is added next? ____________________

(g) What can be added to the yoghurt when it has been produced?____________________

(h) Why do you think dairies might use these additives?____________________

____

5. What is the purpose of the following in the production of beer?

(a) Malting.

____

____

(b) Grinding of the malt.

___

___

(c) Addition of hops.

___

___

(d) Addition of *Saccharomyces cerevisiae*.

___

___

(e) Fermentation.

___

___

6. The graph (Fig. 10.2) shows the relationship between light intensity and the manufacture or utilisation of carbohydrate in two different plants. The graphs use the same scales which have arbitrary units for (i) the use of carbohydrate, and (ii) the light intensity.

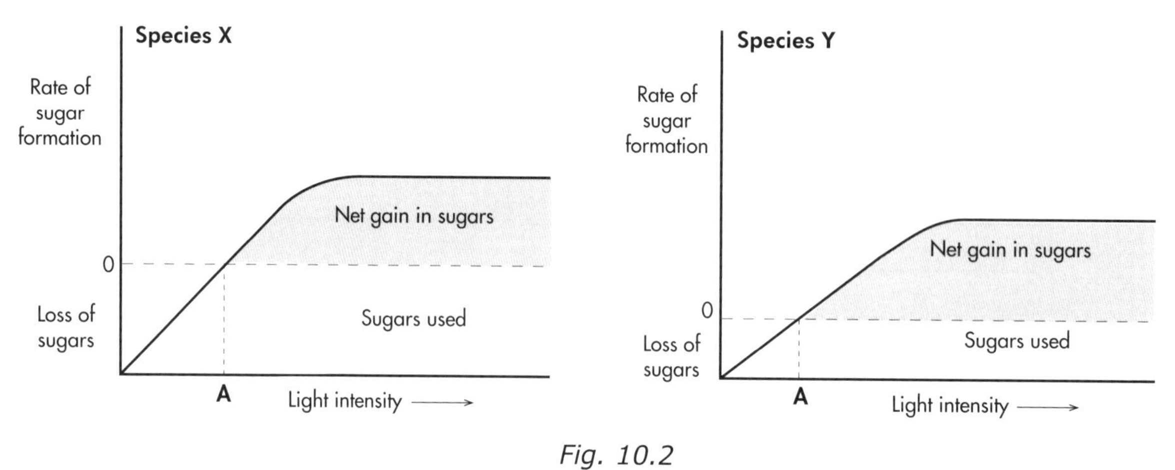

Fig. 10.2

(a) What process is producing the net sugar gain in the plant? ___

(b) Why is the net sugar loss greatest when the light intensity is zero?

___

___

(c) At the point A on each graph there is no net sugar gain or loss. Why do you think this is? ___

___

(d) State with reasons which of the two species you would expect to find in a shady habitat. ___

___

___

(e) Suggest why the rate of net sugar gain levels off after a certain light intensity is reached.

____

____

7. (a) Distinguish between the light-dependent and the light-independent stages of photosynthesis.

____

____

(b) What are the products of the light-dependent stage of photosynthesis?

____

(c) How do the chlorophyll molecules function in photosynthesis?

____

____

(d) How are the end products generated?

____

____

(e) What happens to these end products?

____

____

(f) What occurs in the light-independent stage?

____

____

8. Fill in the blanks in this diagram of respiration (Fig. 10.3).

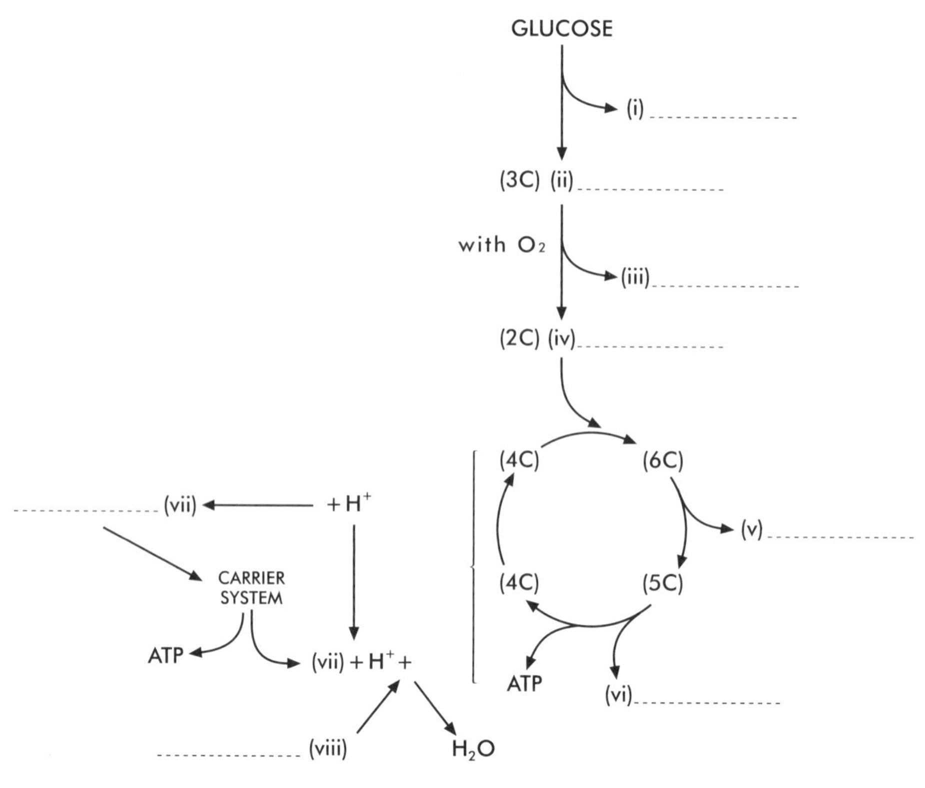

Fig. 10.3

SECTION B

1. (a) In an experiment a leaf is tested for the presence of starch. What would be the purpose of carrying out such a test?

____

(b) How would you test for starch? ______________________________

(c) A variegated leaf (a leaf with white parts to it) was tested for starch after being left in bright sunlight for a few hours. It was found that the green parts of the leaf were stained blue/black and the white parts of the leaf were stained brown (iodine colour).

(i) Give a reason for this result.

____

____

____

(ii) What does this tell you about photosynthesis?

____

____

____

2. The diagram (Fig. 10.4) shows an experiment set up to determine the rate of photosynthesis at different concentrations of CO_2.

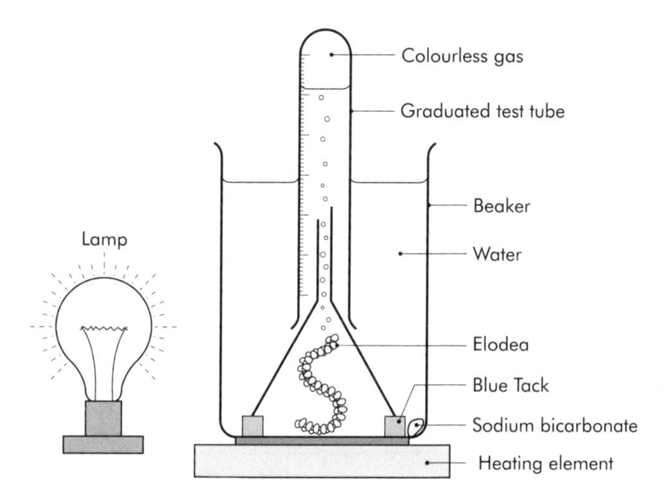

Fig. 10.4

(a) What variables were kept constant in this experiment? ______________________

____

(b) How were they kept constant?

____

____

(c) How was the concentration of CO_2 varied? ______________________

____

(d) How was the rate of photosynthesis measured? ______________________

____

3. (a) In the space provided draw a well-labelled diagram of the apparatus you would set up to show the effect of light intensity on the rate of photosynthesis.

(b) The usual way of measuring the rate of photosynthesis is to count the number of bubbles produced. What errors could there be in this method?

___

___

___

(c) Suggest two ways in which it would be possible to change the level of light?

___

___

(d) Name two factors that need to be kept constant in this experiment and explain how this is achieved.

___

___

(e) On the graph below (Fig. 10.5) draw the result you would expect.

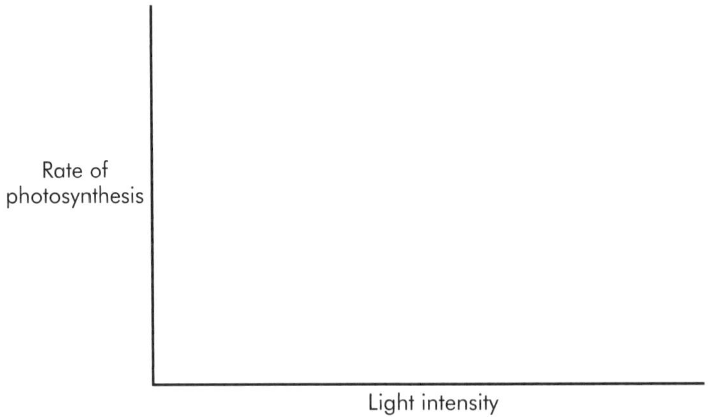

Rate of photosynthesis

Light intensity

Fig. 10.5

SECTION C

1. (a) Explain in simple terms the process of photosynthesis describing where the reactant molecules come from and what happens to the end products.

(b) Describe an experiment to demonstrate the effect of light intensity on photosynthesis and give an indication of the types of results you would expect.

2. (a) Describe the process of aerobic respiration. Explain what happens in humans and microorganisms in the absence of oxygen.

(b) Describe the processes by which humans use fermentation to produce food or drink.

3. Bicarbonate indictor can be used to indicate the change in CO_2 in the air. This indicator will go red when the air from the normal atmosphere is bubbled through it. If the level of CO_2 is reduced in this air then the indicator will go purple. If the level of CO_2 is increased the indicator will go yellow.

In an experiment a number of test tubes, as drawn below (Fig. 10.6), were set up. The indicator was red in all tubes at the start of the experiment. By the end of a few hours the following results were obtained: (A) red, (B) purple, (C) yellow and (D) yellow.

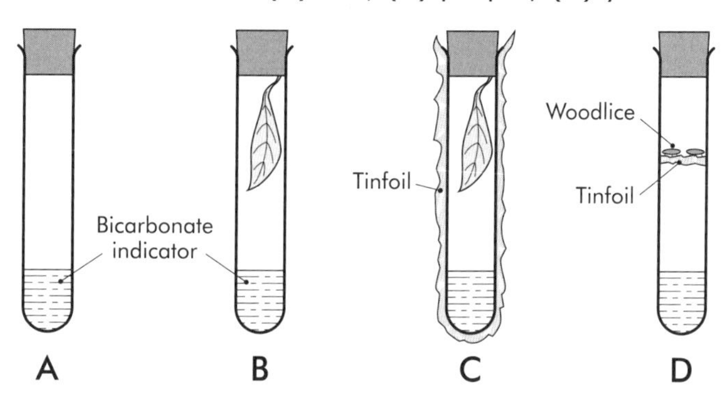

Fig. 10.6

(a) Explain each of the results obtained.

(b) What was the purpose of test tube A?

(c) What is the purpose of the tinfoil around test tube C?

(d) A fifth test tube was used which replaced the tinfoil around the leaf with a piece of muslin that excluded some of the light, and where the bicarbonate indicator was red at the end of the experiment. What happened in this test tube?

4. The process of photosynthesis is composed of two stages, the light-dependent stage and the light-independent stage.

(a) State where these processes take place in the plant cell.

(b) State briefly the meaning of the term light-dependent stage.

(c) Describe how the products of the light-dependent stage are produced.

(d) Explain which of these products are used in the light-independent stage and how they are used.

(e) What are the end products of photosynthesis?

5. The graphs (Fig. 10.7) show the result of experiments to measure the rate of photosynthesis at different light intensities. Answer the following making use of the information on the graphs as appropriate.

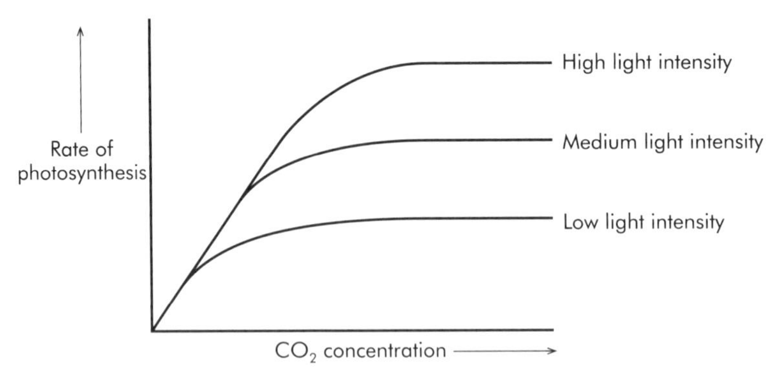

Fig. 10.7

(i) State the relationship between CO_2 concentration and the rate of photosynthesis.

(ii) Explain the effect of varying light intensity on the rate of photosynthesis.

(iii) Why would you keep the temperature constant during this experiment?

6. (a) Describe in detail the process of aerobic respiration.

(b) Describe how you would set up an experiment to demonstrate anaerobic respiration. How would you show the presence of the two end products?

7. (a) Draw a labelled diagram to show the structure of a mitochondrion.

(b) Name the end product of aerobic glycolysis. State the number of carbon atoms in this substance. What is this substance converted to under anaerobic conditions in a muscle cell and in *S. cerevisiae*?

(c) Flowering plants respire during the hours of daylight, but unlike animals, it is difficult to detect the carbon dioxide which they produce. Why is this?

8. The diagram (Fig. 10.8) shows an outline of some of the events in the complete aerobic respiration of glucose in a cell.

(a) Give a balanced chemical equation to summarise the process of aerobic respiration.

(b) Name the polysaccharide storage material in mammals from which the glucose is produced for respiration and give one major storage location in the body.

(c) Give the terms used in each case to describe the series of reactions in stage A and in stage B on the diagram. State where in the cell each of these stages takes place.

(d) Outline the main features of stage B to show the completion of the process of aerobic respiration.

(e) The blood is unable to deliver an adequate supply of oxygen to the muscles during short intensive periods of exercise, e.g. an athlete running a 100 m race. The athlete is said to have an oxygen debt and may suffer muscle pain and cramps as a result. Indicate how the absence of oxygen modifies the process shown in the diagram and suggest why the athlete's pain and cramp disappear a short time after the race.

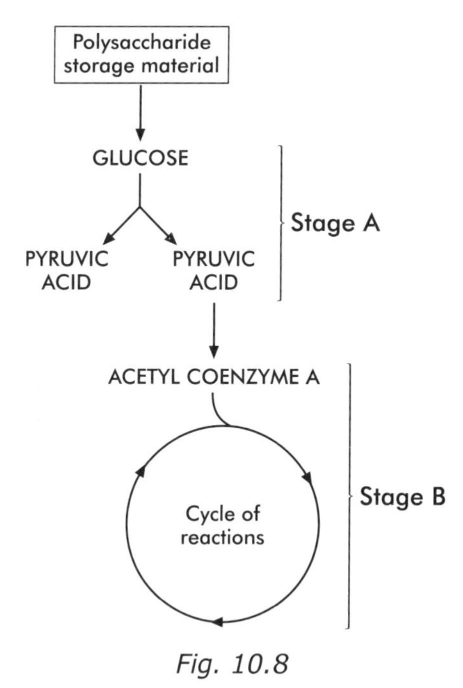

Fig. 10.8

9. The effect of temperature on the rates of (a) apparent photosynthesis (net uptake of CO_2 in light) and (b) respiration (CO_2 produced in the dark) was determined. The results, expressed as milligrams of CO_2 taken up and released, per gram of dry weight of leaf per hour (mg CO_2/g/h), are given below.

Temperature (°C)	7	10	15	19	22	28	31
Uptake – apparent photosynthesis (mg CO_2/g/h)	1.3	2.3	2.8	3.1	2.8	2.5	1.8
Release – respiration (mg CO_2/g/h)	0.3	0.6	0.7	1.2	1.8	2.1	2.7

(a) Why is the term apparent photosynthesis used?

(b) What is the real amount of photosynthesis being carried out in the plant?

(c) Calculate the rates of true photosynthesis at each temperature, assuming that the rate of respiration in the light is equal to the rate of respiration in the dark.

(d) On the same piece of graph paper plot the results for apparent photosynthesis, true photosynthesis and respiration at each temperature.

(e) Explain the relationship between true photosynthesis and temperature.

chapter 11 *Movement through Cell Membranes*

SECTION A

1. (a) Diffusion is________________________________

(b) What gases normally diffuse in and out of cells? ____________________

____

(c) The difference in concentration between two substances is called a

____

(d) What will always take place when there is such a difference in concentration?

____

2. (a) Define osmosis.________________________________

____

(b) In what way can osmosis be considered a special case of diffusion?

____

____

(c) What is meant by the term semi-permeable?

____

____

(d) What are in cell membranes that make them semi-permeable?

____

(e) If these chemicals in a membrane are changed what effect will this have?

____

____

3. (a) If an animal cell is immersed in pure water what will happen and why?

____

____

____

(b) How do single-celled organisms living in fresh water overcome this problem?

____

____

____

(c) What is needed for this process to occur?

____

(d) What would happen to human cells if they were placed in a concentrated salt solution?

____

(e) What does this tell you about the concentration of blood in a human?

____

____

____

(f) What is osmoregulation?

____

____

4. (a) What surrounds a plant cell membrane?

____

(b) In the spaces below draw diagrams of plant cells after they have been placed: (i) in pure water, and (ii) in a strong salt solution.

(i)

(ii)

(c) What are the names given to the condition of each of these cells?

(i) ____________________ (ii) ____________________

(d) How do herbaceous plants make use of (i)?

____

(e) When a plant wilts what has happened to its cells?

____

____

5. (a) 'Food preservation makes use of osmosis.' Explain this statement.

(b) What types of food can be preserved using this method?

(c) What chemicals are added to preserve food in this manner?

(d) What could be the problems for humans in eating a lot of preserved foods?

SECTION B

1. In an experiment potatoes were cut into 'chips', where each was carefully cut until it weighed 10 g. Each of these chips was then put into a sugar solution of differing concentration. At the end of one hour each 'chip' was dried by rolling it in filter paper and re-weighed. The results obtained are in the table below.

Concentration of sugar (moles)	0.0	0.1	0.2	0.3	0.4	0.5	0.6
Finishing weight (g)	10.9	10.8	10.6	10.2	9.8	9.4	9.2

(a) Draw a graph of change in weight against concentration of solution on the grid below (Fig. 11.1).

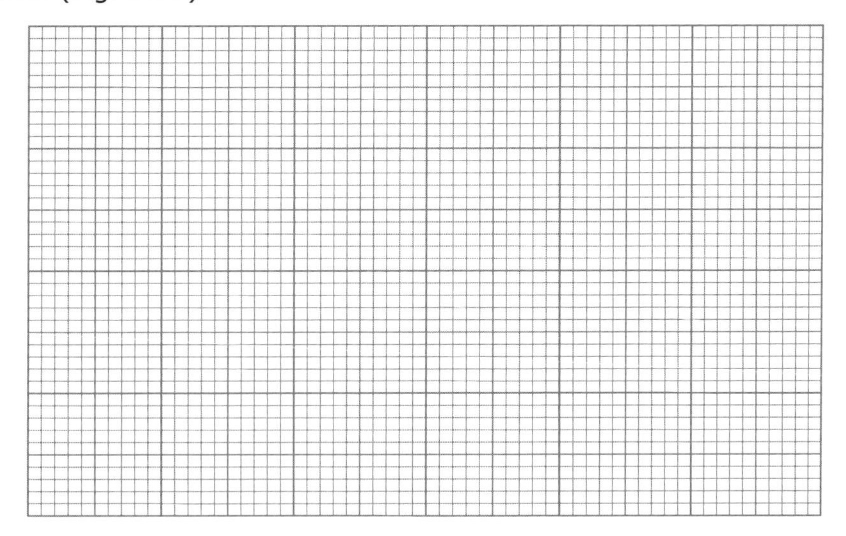

Fig. 11.1

(b) What process is being demonstrated here?

____

(c) From the graph what is the normal concentration of the potato cells? __________

(d) What name is given to the condition of the cells in: (i) the 0.0M solution, and (ii) the 0.6M solution?

(i) ______________ (ii) ______________

(e) What control could you set up to demonstrate that this process only takes place in living cells? ______________________________

(f) Why were the potato 'chips' dried on filter paper before weighing? ______________

____

2. Visking tubing is made of a plastic material that acts as a semi-permeable membrane and only allows small molecules to pass through it. In an experiment four tied pieces of visking tubing were filled with honey or water and placed in different solutions as shown below (Fig. 11.2).

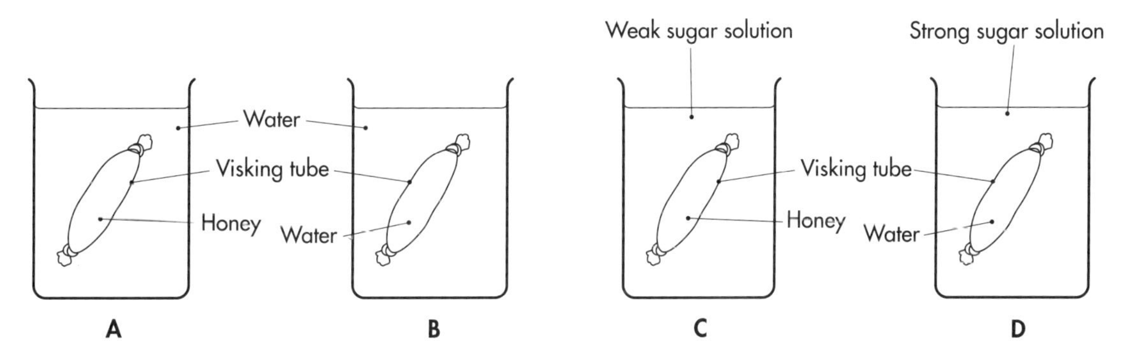

Fig. 11.2

(a) What will happen in each tube after 24 hours?

A ______

B ______

C ______

D ______

(b) Which of the four beakers is the control? ______

(c) Explain the results in beaker A and beaker C.

A

______

______

______

C

______

______

______

SECTION C

1. Comment on the validity of the following statements:

(a) 'Diffusion and osmosis are both passive processes which require no energy input from a cell, but they are both vital for the survival of cells.'

(b) 'Because it is the proteins in a membrane that control the passage of materials, all membranes allow the same chemicals in and out.'

(c) 'Osmoregulation is vital for the survival of multicellular organisms.'

(d) 'Osmosis affects plant and animal cells in the same way.'

(e) 'Water holds all flowering plants upright.'

2. (a) Explain how it is that adding salt or sugar can preserve foods.

(b) Describe using labelled diagrams what happens to a plant cell when it is placed in solutions of differing concentrations.

(c) Describe an experiment to demonstrate that osmosis only occurs in a living cell.

chapter 12 *Cell Continuity*

SECTION A

1. (a) What are the functions of cell division? ____________________

(b) A chromosome is composed of ____________________

(c) When do chromosomes become visible as distinct structures?

(d) Distinguish between haploid and diploid cells.

(e) What is the chromosome number of gametes?

(f) Why is this the case? ____________________

2. (a) What is produced when a cell divides by mitosis?

(b) When do cells undergo mitosis?

(c) How many stages are there in mitosis? ____________________

(d) When does the DNA replicate?

3. In the outline diagram of the cells provided, draw a diagram of each stage of mitosis in the correct order (Fig. 12.1).

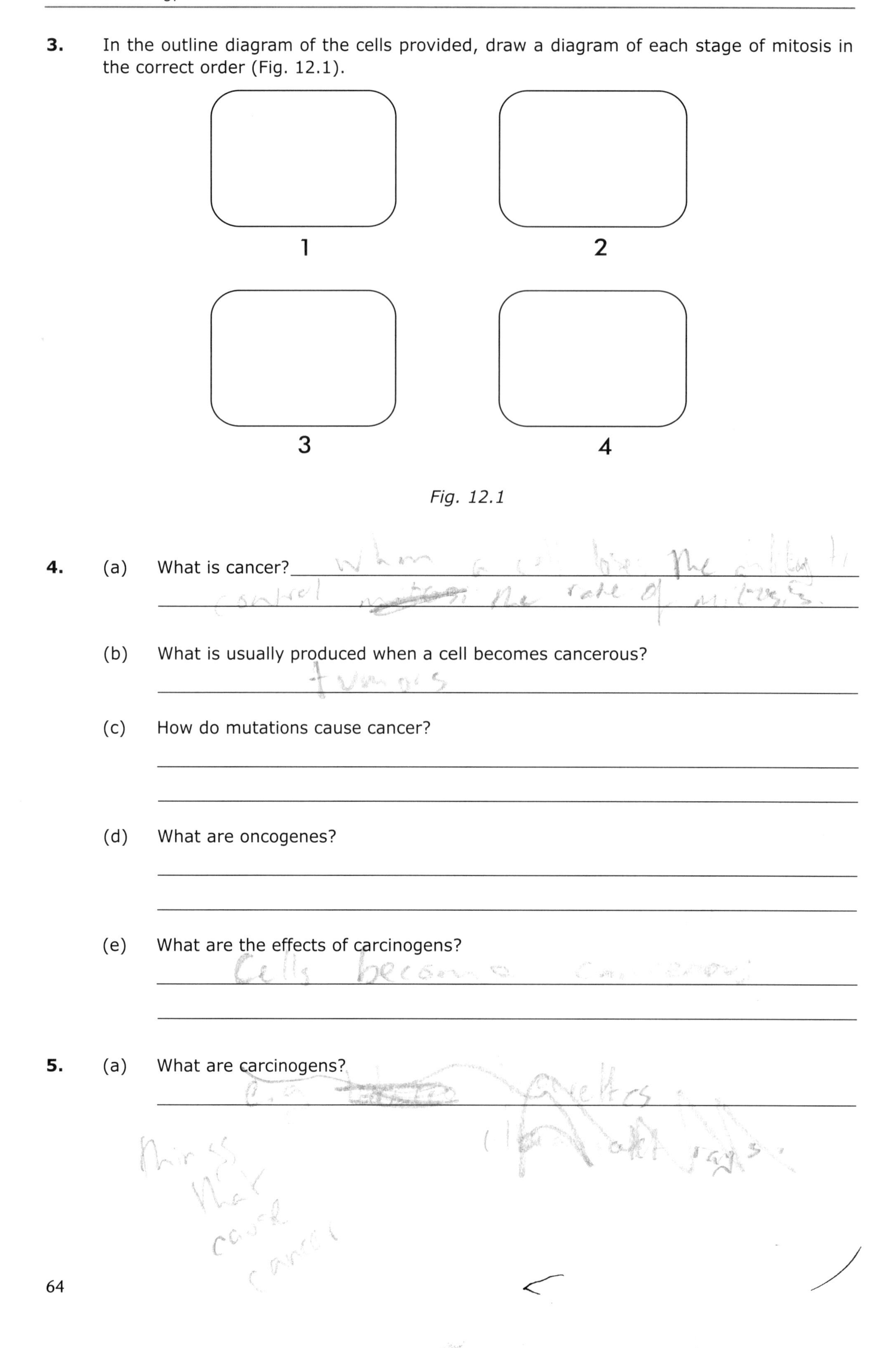

Fig. 12.1

4. (a) What is cancer? ______________________________

____

(b) What is usually produced when a cell becomes cancerous?

____

(c) How do mutations cause cancer?

____

____

(d) What are oncogenes?

____

____

(e) What are the effects of carcinogens?

____

____

5. (a) What are carcinogens?

____

(b) List four carcinogens and explain how their intake can be avoided.

A ____________ ____________________

B ____________ ____________________

C ____________ ____________________

D ____________ ____________________

(c) What else might increase the risk of an individual getting cancer?

____

____

6. (a) Define meiosis. ____________________

____

(b) What type of cell does this process produce?

____

(c) What is the usual function of this cell?

____

(d) What are the benefits of meiosis to an organism?

____

____

7. 'Organisms which reproduce asexually produce clones.'

(a) What can you say about the genetic make-up of such clones?

____

(b) What could be the advantages of clones?

____

____

(c) What might be the disadvantages of clones?

____

____

(d) List some organisms which reproduce in this manner.

____

____

8. (a) Identify the stage of mitosis in the following diagram (Fig. 12.2).

Stage: Anaphase

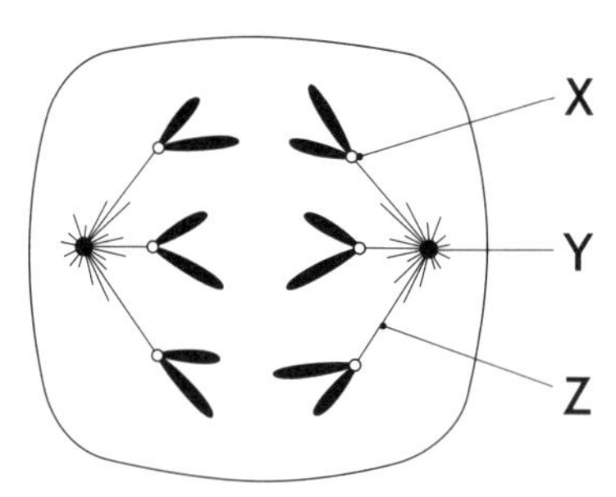

Fig. 12.2

(b) Label the parts X, Y and Z.

X ____________ Y ____________ Z ____________

(c) Describe the process of prophase.

____

____

____

____

(d) In the space below draw a labelled diagram of the second stage of mitosis.

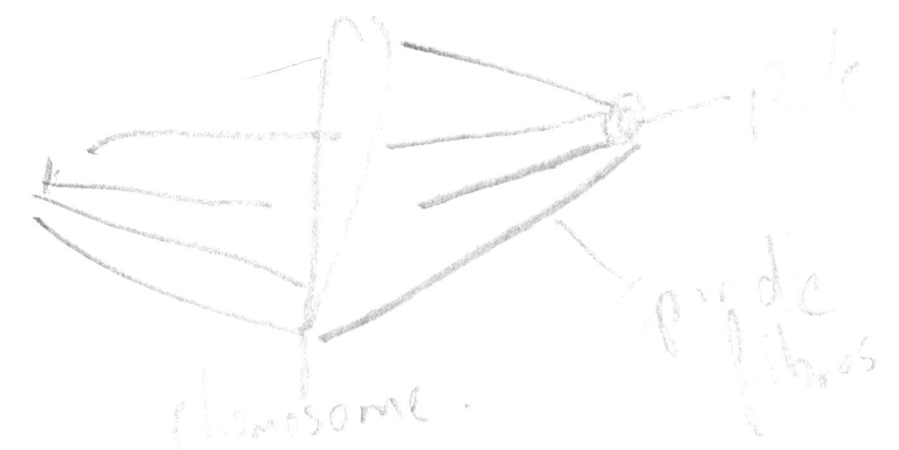

(e) What are the differences in telophase in a plant cell and in an animal cell?

____

____

____

SECTION C

1. (a) Describe the process of cell division using simple labelled diagrams.
(b) What can happen if control of this process is lost?
(c) What are the possible causes of this loss of control?

2. Using well-labelled diagrams describe the process of mitosis in an animal cell. What differences, if any, would be seen in a plant cell undergoing the same process?

chapter 13 *Cell Diversity*

SECTION A

1. Fill in the blank spaces using an appropriate number or from the following words (some words may be used more than once):

organs identical systems tissues multicellular

Most living organisms are made up of many cells and are called ________________. In most organisms all the cells are not ________________ but are grouped together into ________________ made up of identical cells. These ________________ are grouped together to form functioning units called ________________. In animals these group together to form ________________ each of which carry out specific functions. In humans there are ________________ of these ________________, which form the organism.

2. (a) Define tissue. ____

____

(b) List and describe the four groups of tissue in: (i) animals, and (ii) plants (using function as the method of classification).

(i)

1. ____________ ________________________________
2. ____________ ________________________________
3. ____________ ________________________________
4. ____________ ________________________________

(ii)

1. ____________ ________________________________
2. ____________ ________________________________
3. ____________ ________________________________
4. ____________ ________________________________

3. This is a diagram of an animal tissue (Fig. 13.1).

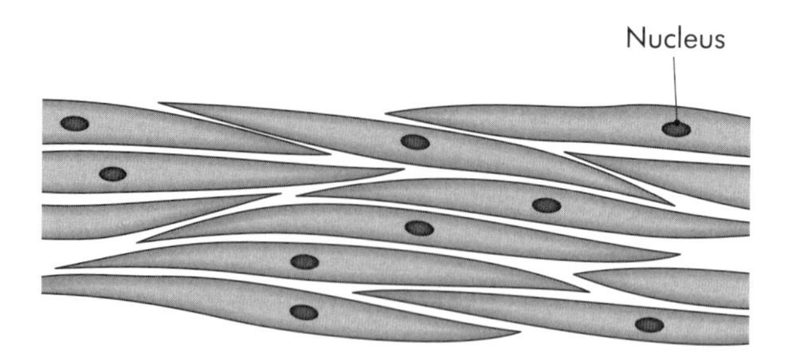

Fig. 13.1

(a) Name this tissue. ______________________________

(b) Name two other similar types of tissue. ______________________________

(c) Compare these three tissues.

____

____

____

____

4. (a) In plants what is the dermal tissue? ______________________________

(b) Describe the dermal tissue of herbaceous plants.

____

(c) What is the function of guard cells in the epidermis?

____

____

____

____

(d) What are meristematic cells? ______________________________

(e) Compare these two types of cells. ______________________________

____

5. (a) What are tissue cultures? ______________________________

(b) What is the genetic make-up of the cells in a tissue culture?

____

(c) What are the possible benefits to humans of tissue cultures?

____

(d) What might be the problems in trying to produce organs in this manner?

____

____

6. (a) What is an organ? ____

____

(b) List the six plant organs.

(i) ________________ (ii) ________________ (iii) ________________
(iv) ________________ (v) ________________ (vi) ________________

(c) What are organ systems?____

(d) Why are organ systems found only in animals?

____

____

7. (a) Define a species. ____

(b) Organisms are classified by grouping them together. They are grouped together in decreasing order of similarity. Starting with species name the main levels used to classify.

Species __________ __________ __________ __________ __________ Kingdom

(c) What two groupings are used to name any organism?

____

(d) Give any one example of this binomial system. ________________________

(e) How many kingdoms do scientists most commonly use? ____________________

(f) List these kingdoms. ____

____

chapter 14 *Genetics and Evolution 1*

SECTION A

1. (a) A section of DNA that codes for a particular trait is called a ____________________

(b) Define allele. ____

(c) A DNA strand with attached proteins is called a ____________________

(d) A diploid cell contains ____________________________________

(e) A haploid cell contains ____________________________________

2. (a) Label the following diagram of DNA (Fig. 14.1).

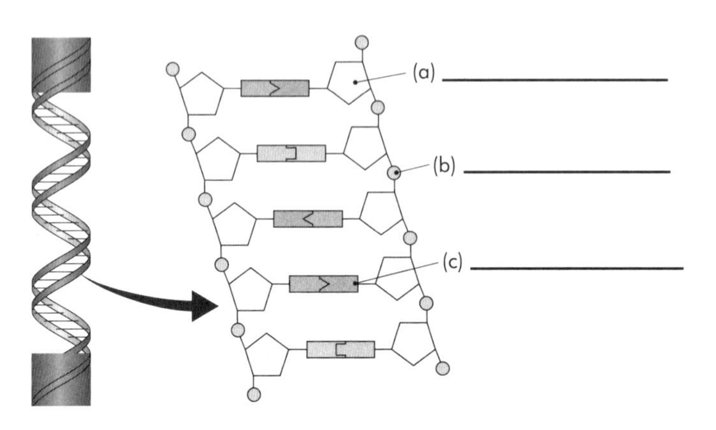

Fig. 14.1

(b) The bases in DNA are always found in pairs A with ____________ and C with ____________

(c) In genes the order of the bases along the DNA chain codes for

____

(d) The non-coding sections of DNA are called____________________________

(e) These sections of DNA are important for what identification process?

____

3. In the production of a DNA profile the following procedures can be carried out. Give the purpose of each procedure.

(a) A sample of blood, semen or cheek cells is taken.

____

____

(b) Enzymes are added to the sample.

____

____

(c) The DNA is added to a gel and an electric current is applied to the gel.

____

____

(d) A nylon membrane is placed over the gel.

____

____

(e) Radioactive probes are added to the nylon.

____

____

(f) The nylon is placed in contact with X-ray film.

____

____

4. (a) What is genetic screening?

____

____

(b) Give two other uses for a DNA profile. ______________________________

(c) What are the benefits of genetic screening?

____

____

____

(d) What are the ethical problems associated with genetic screening?

____

____

____

5. (a) Why is it essential that the DNA can make an exact copy of itself?

____

____

(b) What chemicals control the process of DNA replication in cells?______________

(c) What is pulled apart in the first stage of replication? ______________

(d) What is matched together in this stage?______________

(e) Where are the free bases found? ______________

(f) What results from this process?______________

(g) What consequence does this have for the two new DNA strands?

____

____

6. (a) How does the DNA control the activity of the cell? ______________________

(b) What determines the type of protein that is made?______________________

(c) How many bases make up the genetic code?______________________

(d) What is the first step in the process of protein manufacture?______________________

(e) Give one difference between mRNA and DNA.______________________

(f) The mRNA is complimentary to the DNA. What does this mean?

____

____

(g) Where is the protein made? ______________________

(h) What happens to the newly manufactured chain of amino acids? ______________________

7. Define the following terms used in genetics:

(a) Heterozygous ______________________

(b) Homozygous ______________________

(c) Dominant gene ______________________

(d) Recessive gene ______________________

(e) Genotype ______________________

(f) Phenotype ______________________

8. In the following genotypes, state whether it is homozygous or heterozygous and give all the possible gametes that could be produced and the proportion in which they will be found.

(a) RR ______________________ ____________ ____________

(b) Aa ______________________ ____________ ____________

(c) Hh ______________________ ____________ ____________

(d) Rr ______________________ ____________ ____________

9. In peas, green seed colour is dominant to yellow seed colour. A plant that produces yellow seeds is crossed with a plant homozygous for green colour.

(a) What is the genotype of the yellow plant? ______________________

(b) What gametes can this plant produce? ______________________

(c) What is the genotype of the green plant? ______________________

(d) What gametes can it produce? ____________________

(e) What will be the genotype of the offspring? ____________________

(f) What will be the phenotype of the offspring? ____________________

(g) What possible gametes could this offspring produce? ____________________

10. A grey rat homozygous for coat colour was crossed with a black rat homozygous for coat colour. All the offspring had grey coats.

(a) What is the dominant gene for colour? ____________________

(b) How do you know this? ____________________

____

(c) What is the recessive gene for colour? ____________________

(d) How do you know this? ____________________

____

(e) Give the genotypes of both parents.

____

(f) Give the possible gametes produced by each parent.

____

(g) Give the genotype of the offspring. ____________________

11. In horses trotter (T) is dominant to pacer (t). A trotter stallion is mated on a number of occasions to the same mare. On each occasion the resultant colt was a trotter. In the space provided below draw a labelled diagram showing this genetic cross.

12. In guinea pigs the gene for long hair is recessive to short hair. A long-haired male was mated with a heterozygous short-haired female.

(a) What are the genotypes of the two adults? ____________ ____________

(b) What gametes can they produce? ____________ ____________

(c) What are the genotypes of the offspring? ______

(d) What are the phenotypes of these offspring? ______

(e) In what proportion will they be produced? ______

13. (a) In some variety of cats white fur is dominant to black fur. Two white cats bred and produced both white and black kittens.

(i) What is the genotype of the adult cats? ______ ______

(ii) How do you know this? ______ ______

______

(iii) Would you expect there to be an equal number of black and white kittens?

______

(iv) Which colour would be more common? ______

(b) A black cat bred with a white cat and produced an equal number of black kittens and white kittens.

(i) What would be the genotype of the white parent? ______

(ii) How do you know this? ______

______

(iii) How do you know the genotype of the black adult? ______

______

(iv) In the space below draw a diagram of the cross.

14. In short-horn cattle the coat colour red is codominant with white giving a roan colour in the heterozygous condition.

(a) What is meant by the term codominant? ______

______

(b) A red bull is bred with a white cow.

(i) What is the genotype of the cow? ______

(ii) What is the genotype of the bull? ______

(iii) What is the genotype and phenotype of the offspring? ______

15. In Andalusian fowl, the heterozygous condition of the allele for black plumage and white plumage is blue. In the space below show the possible crosses between a blue-plumed fowl with birds of the following plumage: (a) white, (b) blue and (c) black.

(a)

(b)

(c)

16. (a) What are the differences between the autosomes and the sex chromosomes?

___

___

(b) What are the two types of sex chromosomes? ___ ___

(c) What type of sex chromosomes do (i) males and (ii) females have?

(i) ___ (ii) ___

(d) In what respect can it be said that females produce only one type of gamete?

___

___

(e) Why can it be said that the father determines the sex of the child?

___

___

___

17. An ultracentrifuge is a laboratory device that spins fluid-filled tubes at very high speeds. This will separate out particles suspended in the fluid depending on their weight. The heavier particles are found further down the tube. When this was done to tubes containing sperm it was discovered that two bands of sperm were produced.

(a) What is the difference between these two types of sperm? ____________________

____

(b) Why is one type of sperm heavier than the other? ____________________

____

(c) What is in the sperm found nearer the top of the tube? ____________________

(d) If only the sperm nearest the top of the tube are used in artificial insemination, what would be the most likely sex of the offspring? ____________________

(e) Can you think of any benefits or disadvantages of this procedure?

____

____

____

(f) On the diagram below place the letters X and Y to show how sex is inherited in humans.

	Male	**Female**
Sex chromosomes		
Possible gametes		
Possible offspring		

18. (a) What are the two types of variation? ______________ ______________

(b) Which of these is important in evolutionary terms? ____________________

(c) Why is this the case? ____________________

____

(d) How does sexual reproduction cause variation? ____________________

____

(e) What else can cause variation? ____________________

19. (a) Give an example of a gene mutation. ____________________

(b) Describe how this causes its effect?

____

____

____

(c) Explain what the term 'carrier' means in relation to this gene.

(d) Give an example of another type of mutation.

(e) Describe how this type of mutation has its effect.

20. (a) What is genetic engineering?

(b) What is the first step necessary to carry out this process?

(c) What is a plasmid?

(d) In what way is the plasmid modified to transfer a gene?

(e) The modified plasmid is described as a vector for the gene. Why is this the case?

(f) What is a transgenic organism?

(g) Give one example of a transgenic organism and give a possible benefit for it.

21. (a) Define evolution.

(b) Who is the person most associated with the theory of evolution?

(c) What is a homologous structure in evolution?

(d) How is a pentadactyl limb a homologous structure?

(e) How is this used as evidence for evolution?

(f) What is natural selection? ______________________________

____

(g) Describe an example of natural selection. ______________________________

____

SECTION B

1. Give scientific explanations for each of the following steps taken in the activity to separate DNA from onion cells.

(a) Add the chopped onion to a salt solution containing washing-up liquid.

____

____

____

(b) Stand the beaker of tissue in a water bath at 60 °C for 15 minutes.

____

____

(c) Place the tissue sample in ice. ______________________________

____

(d) Blend the mixture for only five seconds. ______________________________

____

(e) Filter the blended mixture. ______________________________

____

(f) Add a small amount of protease to some of the filtered mixture.

____

____

(g) Carefully pour ice-cold ethanol on top of the mixture.

____

____

SECTION C

1. (a) Describe the structure of DNA.
(b) Describe the steps taken in the production of a DNA profile for an organism.
(c) What are the possible uses for a DNA profile?
(d) Can you give any of the possible problems associated with this technique?

2. (a) Describe the process by which sex is inherited in humans.
(b) Explain why this process gives an equal number of male and female children.
(c) What role do genes play in heredity?
(d) Explain the difference between genotype and phenotype.

3. In antirrhinum plants the gene for red (C^R) and white (C^W) flower colour demonstrates codominance. When a plant has both genes present (C^RC^W) the flower colour is pink. Using diagrams show the following:

(a) A cross between a red-flowered plant and a white-flowered plant.
(b) A cross between two pink-flowered plants.
(c) A cross between a pink-flowered plant and a white-flowered plant.

If a gardener planted 50 seeds produced from a cross between two plants and all the plants produced had pink flowers, what could you say about the likely genotypes and phenotypes of the original plants?

4. In humans the gene for brown eyes (B) is dominant to the gene for blue eyes (b).
(a) A brown-eyed woman and a blue-eyed man have six children all of whom have brown eyes. Using the information given, explain the likely genotype of both parents and of their children.
(b) A blue-eyed woman and a brown-eyed man have two children both of whom have blue eyes. Using a diagram explain this cross.
(c) Two brown-eyed parents produce a child who has blue eyes. What is the genotype of these three individuals?

5. (a) How is variation important in evolution?
(b) What are the causes of genetic variation? Give an example in each case and describe how these may affect humans.

6. (a) Describe the process of genetic engineering using a named example.
(b) 'Genetic engineering is simply a modern example of a process as old as the first farmers.' Discuss the validity of this statement.

7. (a) What is evolution?
(b) Describe three pieces of evidence which can be used to support this theory.
(c) What is natural selection and how does it affect populations?

8. (a) Describe an experiment used to extract DNA from a tissue sample.
(b) Give some possible uses of this technique in biology.

chapter 15 *Genetics 2 (Higher Level Only)*

SECTION A

1. (a) Who was Gregor Mendel? ______________________________

(b) What was different in his approach to his work?

____

(c) What are pure-breeding plants? ______________________________

(d) Why was it important that Mendel used pure-breeding plants in his experiments?

____

____

____

(e) What did the letters P, F_1 and F_2 mean when used by Mendel in his results?

P ______________ F_1 ______________ F_2 ______________

2. (a) What did Mendel conclude from his experiments? ______________________________

____

____

(b) What was Mendel's first law? ______________________________

____

____

(c) Give the modern explanation of Mendel's first law. ______________________________

____

____

(d) What is the difference between a probability and a certainty?

____

____

____

3. In peas green seed is dominant to yellow seed. A homozygous green-seeded plant is crossed with a homozygous yellow-seeded plant. Show the phenotypes produced in this cross and in the following F_2 generation.

Parents ______________ X ______________

Gametes ______________ ______________

F_1 Genotype ______________ X ______________

Phenotype ______________

Gametes ______________ ______________

F_2 Genotypes ______________________

Phenotypes ______________________

4. (a) What is the modern word used to describe the term 'factor' as used by Mendel?

____

(b) Where can such factors be found in a cell?______________________________

(c) Differentiate between the following pairs of terms.

(i) Homozygous and heterozygous.

____

____

(ii) Dominant and recessive.

____

____

(iii) Haploid and diploid.

____

____

5. In humans the brown-eyed gene (B) is dominant to the blue-eyed gene (b). If two parents are heterozygous for the trait, show:

(a) The genotype of the parents. ______________________________

(b) The genotypes of the gametes. ______________________________

(c) The possible genotypes of their children. ______________________________

(d) The possible phenotypes of their children. ______________________________

6. In short-horn cattle the gene (P) for polled (non-horned) is dominant to the gene (p) for horned. A cow and bull heterozygous for the condition mate over a number of years and produce calves.

(a) What is the probability that the first calf would be horned? ______________

(b) What is the probability that the second calf would be polled? ______________

(c) What is the probability that the second calf would be horned? ______________

(d) Is the answer in (c) different from that in (a)? Why?______________________

____

____

(e) Is it possible that all the offspring produced by these two animals would be horned? Why?

____

____

____

(f) Is it probable that all the offspring would be horned? Why?

____

____

7. A tall tomato plant with entire leaves was crossed with a short tomato plant with divided leaves. The seeds produced were planted and all of the resulting tomato plants were tall with divided leaves. These plants were crossed with themselves and the resultant seeds produced the following types of plants:

(i) 608 tall plants with divided leaves
(ii) 205 tall plants with entire leaves
(iii) 213 short plants with divided leaves
(iv) 70 short plants with entire leaves.

(a) What is a dihybrid cross? ______________________________

(b) Using the information given above fill in the following diagram.

Parents' genotype ________ X ________
Gametes ________ ________
F_1 Genotype ________ X ________ (selfed)
Gametes ________ ________ ________ ________
________ ________ ________ ________

Gametes				

F_2 phenotypes of offspring with ratios:

____

8. (a) What is Mendel's second law?______________________________

____

____

(b) Under what circumstances will this law not hold true?

____

(c) Why is this the case?

____

____

____

(d) Tall plants with red flowers were crossed with short, white-flowered plants. The F_1 generation were all tall red-flowered plants. One of the F_1 plants was crossed with a short white-flowered plant. Some 50 per cent of the offspring were tall and red while 50 per cent were short and white.

(i) Does this cross demonstrate Mendel's second law? ______________________

(ii) Why is this the case?

____

____

____

9. In the following pedigree, four generations of a family are shown indicating the inheritance of colour blindness (Fig. 15.1).

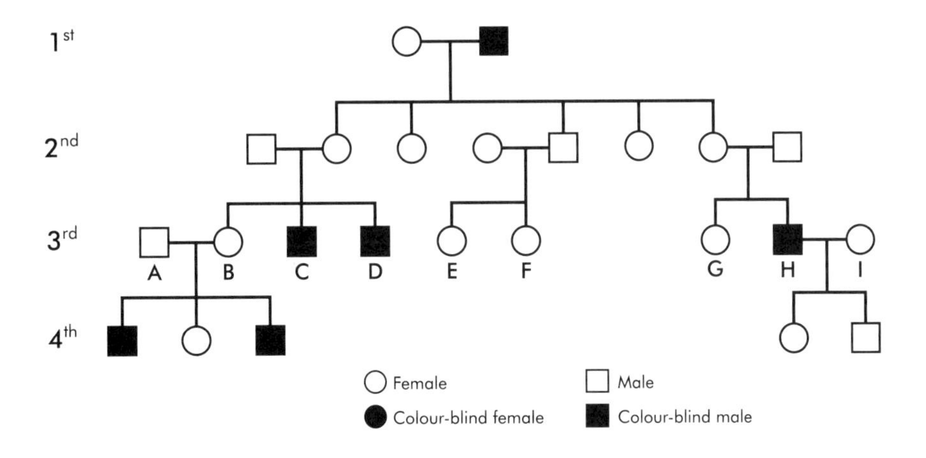

Fig. 15.1

(a) What is meant by the term 'a sex-linked gene'? ______________________

____

(b) In what way does the inheritance of these genes differ from Mendel's laws?

____

____

____

(c) Why do you think that none of the second-generation family is colour blind?

____

____

(d) What is the genotype of the females of the second generation?

____

(e) How is it that in the third generation the male child C is colour blind?______________

____

(f) What is the genotype of the wife of the male H?

____

(g) On what evidence can you give your answer to question (f)?

____

____

10. (a) What is non-nuclear inheritance? ______________________

(b) What is inherited from the mother with the nucleus of the gamete?

____

(c) Why is non-nuclear inheritance largely limited to mitochondria in animal cells?

____

(d) What other cell structure is involved in plant cells? ______________________

____

(e) Certain populations can be identified by groups of genes called markers found in their DNA. By looking at non-nuclear inheritance in Iceland it has been discovered that about 50 per cent of women who founded the population there were Irish.

From your knowledge of genetics explain how it might be possible to state this.

____

____

____

____

11. (a) What is the basic unit that is built into DNA? ______________________

(b) What three chemicals make up this basic unit?

______________ ______________ ______________

(c) What do the following letters stand for in DNA?

A ____________ T ____________ G ____________ C ____________

(d) What is a complementary base pair? ______________________

____

(e) What are the complementary base pairs in DNA?

____

12. (a) Label the following parts of a DNA molecule.

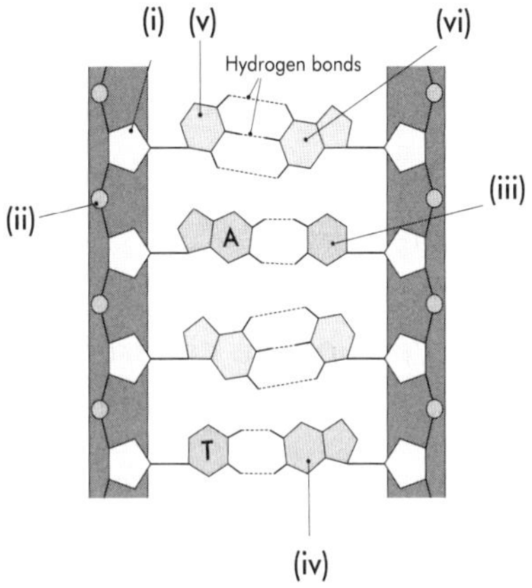

(i) (v) (vi)
Hydrogen bonds
(ii)
(iii)
A
T
(iv)

(i) ______
(ii) ______
(iii) ______
(iv) ______
(v) ______
(vi) ______

(b) What is the normal shape of the DNA molecule? ______

(c) What is the function of the DNA molecule? ______

(d) Why is the order of the bases in the DNA important?

(e) How many letters are in the DNA code? ______

(f) Why is the DNA code described as a triplet code? ______

13. (a) What is mRNA? ______

(b) What is transcription? ______

(c) What is the function of RNA polymerase? ______

(d) What can the mRNA do that DNA cannot? ______

(e) What is the function of a ribosome? ______

(f) What happens in the process of translation?

SECTION C

1. In humans freckles are dominant over no freckles. A man and a woman both freckled have two children both of whom have no freckles.

(a) What are the genotypes and phenotypes of the parents?
(b) What gametes could they possibly have produced?
(c) What are the possible genotypes and phenotypes of the children?
(d) What was the chance that the first child could have had freckles?

2. A black mouse from a pure-breeding strain was crossed with a mouse from a pure-breeding strain of albino mice. All the resultant F_1 mice were black. These F_1 mice were crossed with themselves and they produced offspring, some of which were black and some of which were albino.

(a) Using a diagram illustrate this cross.
(b) What ratio would you expect between the black and albino offspring in the F_2 generation?
(c) Two of the black mice, A and B, from the F_2 generation were crossed with an albino mouse. The mouse A produced only black offspring while mouse B produced some albino and some black mice. Explain these results.

3. (a) What is Mendel's Law of Independent Assortment? What experiment did Mendel carry out to determine this law?
(b) In pea plants the gene for terminal flowers (A) is dominant to the gene for axial flowers (a). A tall pea plant with terminal flowers is crossed with a dwarf plant with axial flowers. The seeds produced were planted and 98 flowers grew: 53 were tall plants with terminal flowers and 45 were tall plants with axial flowers. Explain these results using diagrams where appropriate.

4. A pure-breeding strain of red snapdragon plants was crossed with a pure-breeding strain of white plants. All the resultant plants were pink.

(a) Explain these results. Do these results confirm Mendel's conclusions? What is the name given to this type of gene?

(b) Two of the pink-flowered plants were crossed. Give the genotypes and phenotypes of the offspring with the ratios of the different types of offspring produced.

5. In cats the allele for black and white fur (X^B) is codominant with the allele for ginger and white fur (X^G). This gives a cat that is black, white and ginger (a tortoiseshell cat) (X^GX^B) in the heterozygous condition. This gene is found on the X chromosome.

(a) Using diagrams show the following crosses:
(i) A black and white tom is crossed with a ginger and white queen.
(ii) A ginger and white tom is crossed with a tortoiseshell queen.

(b) Why is it not possible to have a tortoiseshell tom?

(c) This gene exhibits a second exception to Mendel's laws. What is this exception called?

6. In fruit flies eye colour is a sex-linked gene. The allele for red eyes (X^R) is dominant to the allele for white eye (X^r).

(a) Give the results from the following crosses. Include the genotype, phenotype and the sex of the offspring.

(i) A heterozygous red-eyed female and a white-eyed male.

(ii) A white-eyed female and a red-eyed male.

(b) Explain why it is not necessary to state if the white female or any male is heterozygous or homozygous for the gene.

7. (a) Describe the structure of DNA.

(b) Explain how the DNA message can be used to produce protein.

UNIT 3 chapter 16 *Diversity of Organisms*

SECTION A

1. List the five kingdoms and give a simple description, based on nutrition, of the members of each kingdom.

____________	______________________________
____________	______________________________
____________	______________________________
____________	______________________________
____________	______________________________

2. Fill in the following passage using these words:

cell wall spores plasmids spiral chains shape pairs drug resistance temperature membrane capsule flagella groups organelles prokaryote (Monera)

Bacteria belong to the ______________ Kingdom. They can be divided into three groups by using ______________. Spherical bacteria can be found either singly, in ______________ or as ______________. Rod-like bacteria can have ______________ for movement. ______________ bacteria are the third type. All bacteria will have a ______________ on the outside and some, particularly pathogens, have a layer called a ______________ outside this again. Internally, bacteria are very simple and contain no ______________-bound ______________. Many bacteria contain extra chromosomal DNA called ______________. A number of these structures contain genes for ______________. In unfavourable conditions some bacteria will produce ______________ which can only be killed by high ______________.

3. (a) List the two types of autotrophic bacteria and describe how they produce their food.

______________ ____

______________ ____

(b) What is a saprophyte? ____

(c) List the three groups of symbiotic bacteria and explain what each term means.

______________ ____

______________ ____

______________ ____

(d) What is an obligate anaerobe? ______________________________________

(e) What might be the advantage to a bacterium, if it could undertake both types of respiration?

____

____

(f) An anti-tetanus injection is given to a patient only after a deep wound (particularly a puncture wound), but not if the wound is shallow. What does this tell you about the bacterium that causes this disease?

____

____

4. (a) What is an antibiotic? ____

(b) Why are antibiotics useful in medicine?

____

____

(c) Where do antibiotics come from? ______________________________________

(d) Why do you think antibiotics are produced in nature?

____

____

(e) What is the problem associated with the overuse of antibiotics?

____

____

____

(f) Can you suggest ways in which these problems can be minimised?

____

____

____

(g) Why are antibiotics not prescribed for someone suffering from the flu or the common cold?

____

____

5. (a) In the space below draw a labelled diagram of the fungus *Rhizopus*.

(b) How does *Rhizopus* get its nutrition?

____

____

(c) (i) What are sporangiophores? ______________________

(ii) What is their function? ______________________

(d) What is the difference between the positive (+) strain and the negative (–) strain of *Rhizopus*?

____

____

(e) How is a zygospore produced? ______________________

____

(f) What are the advantages to *Rhizopus* in this form of reproduction?

____

____

____

6. (a) In what way is *Saccharomyces* atypical of the Fungus Kingdom?

____

____

(b) Asexual reproduction in *Saccharomyces* can produce a growth known as a yeast. What is this?

______

______

(c) What are the main economic uses of *Saccharomyces cerevisiae?*

______

(d) It has been said that the use of *Saccharomyces* is the first example of biotechnology. Why do you think this is the case and would you agree?

______

______

______

(e) Why is it that the economic uses of *Saccharomyces* require the exclusion of oxygen?

______

______

7. (a) What is the difference between the principle of sterility and that of asepsis?

______

______

______

(b) Why might the principal of asepsis be used?

______

(c) Give an example of the use of asepsis.

______

(d) Why do many scientists work in laboratories with negative pressure to the outside?

______

______

(e) Why are all waste materials sterilised before leaving such a laboratory?

______

______

8. (a) Label the following diagram of an *Amoeba* (Fig. 16.1).

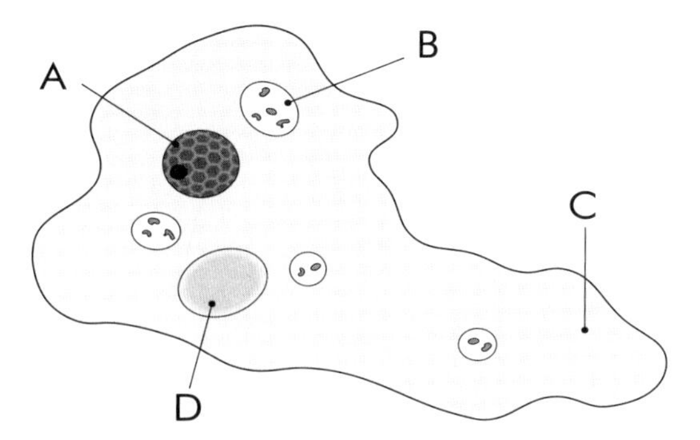

Fig. 16.1

A ____________ B ____________ C ____________ D ____________

(b) What is the purpose of the contractile vacuole?

____

(c) What would *Amoeba* require for a contractile vacuole to function?

____

(d) Why would this be the case?

____

____

(e) To what kingdom does *Amoeba* belong? ____________________

9. (a) In what way do the cells of prokaryotes differ from those of eukaryotes?

____

(b) Why might eukaryotes be considered as more advanced cells than prokaryotes?

____

____

____

(c) What is present in mitochondria that might suggest that they were once free-living cells?

____

(d) What are the differences between a nucleus and a nucleoid?

____

____

10. (a) The graph below shows a growth curve of a bacterium. Name each of the four stages labelled and describe what is happening at each stage (Fig. 16.2).

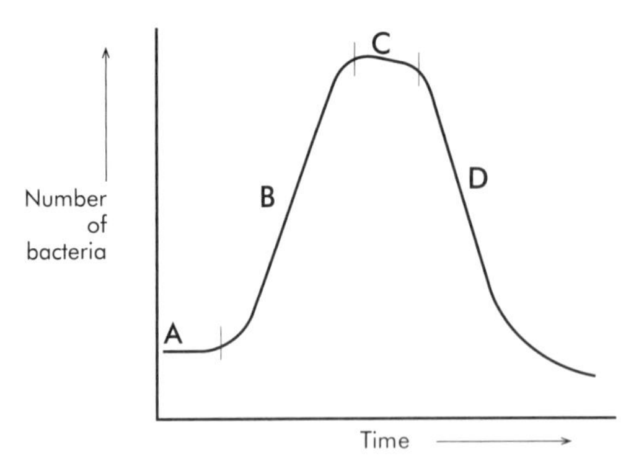

Fig. 16.2

A ____

____

____

B ____

____

____

C ____

____

____

D ____

____

____

(b) When is a continuous method of production of microbes used in industrial applications?

____

____

(c) When is a batch method of production used?

____

____

SECTION B

1. Explain why each of the following procedures was undertaken in the experiment looking at the growth of leaf yeasts.

(a) The bench was wiped with a disinfectant solution.

____

____

(b) The leaf discs were attached to the lid of the petri dish with petroleum jelly.

____

____

(c) One agar plate was unopened.

____

____

(d) The plates were kept, lid uppermost, at room temperature for 24 hours.

____

____

(e) The agar plates were turned upside down after this and left at room temperature.

____

____

(f) The plates were examined regularly.

____

____

(g) At the end of the experiment the agar plates were put into an autoclave.

____

____

SECTION C

1. (a) What are the benefits of classifying living organisms?

(b) Give a brief description, with examples, of the most commonly accepted classification system in use today.

(c) Why do you think scientists changed from a two-kingdom classification system?

2. Give a description of the organisms found in the Prokaryote (Monera) Kingdom under the following headings: (a) cell structure, (b) reproduction and (c) nutrition.

What are antibiotics and what are their uses?

'Antibiotic resistance is a natural phenomenon.' Discuss this statement.
What are the benefits of antibiotics and what are the possible drawbacks of their use?

3. Compare *Rhizopus* and *Saccharomyces* under the following headings: (a) structure, (b) reproduction and (c) economic importance.

Describe an experiment to demonstrate the growth of fungi. Explain the safety precautions that should be taken in this experiment.

The term 'yeast', although commonly used, is not a correct name for the organism *S. cerevisiae*. Explain why this is the case.

4. When working with micro-organisms and in medicine it is important that the risk of contamination by unwanted microbes is avoided. Describe a number of ways in which this can be achieved.

5. Compare the cell structure of a named member of the Protista (Protoctist) Kingdom with that of a member of the Prokaryote (Monera) Kingdom.

6. Draw a graph to illustrate the normal growth curve of a micro-organism and describe the various stages illustrated by the graph.

7. Single Cell Protein (SCP) is usually produced by a continuous flow method but wine is produced in a batch method. What is the difference between these two processes and suggest why a different process is used to produce SCP and wine?

chapter 17 *The Structure of Flowering Plants*

SECTION A

1. The diagram shows a typical flowering plant (Fig. 17.1). Label the parts A–H.

A ____________________

B ____________________

C ____________________

D ____________________

E ____________________

F ____________________

G ____________________

H ____________________

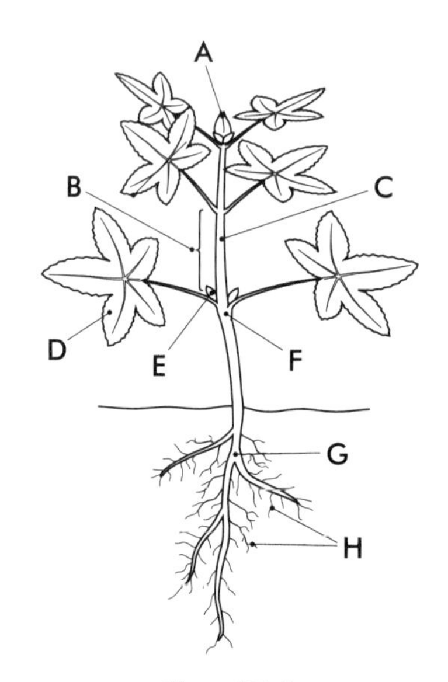

Fig. 17.1

2. From the list below select, in each case, one term which fits the description:

leaf apical bud flower stem root axillary bud
root hairs shoot petiole lateral bud node

(a) Flowering plants are made up of two main parts, the root and the ____________

(b) The part of the plant that is below the ground is the ____________________

(c) Absorption of water is carried out by ______________________________

(d) A leaf stalk is called a ____________________________________

(e) Growth of the stem takes place here ______________________________

(f) A part of the plant involved in sexual reproduction ____________________

3. Distinguish between the following pairs of terms by referring to structure, location or function.

(a) The shoot and the root. ______________________________________

____

(b) Fibrous and tap roots. ______________________________________

____

(c) Apical and axillary buds. ____________________________________

____

(d) Sessile and petiolated leaves. _________________________________

____

4. Match the following plant parts with their correct function(s).

Plant part	**Function**
	Produces seeds
Root	Makes food
Stem	Transports materials between parts of the plant
Leaf	Stores food
Flower	Absorbs water
Bud	Bears flowers, leaves and buds
	Produces new growth

5. Identify each of the three types of cell shown below (Fig. 17.2). State a location and function for each cell type in a plant.

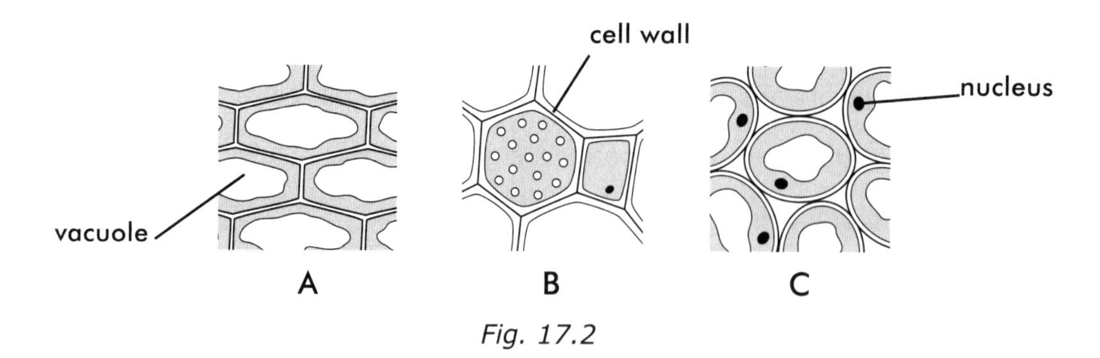

Fig. 17.2

Type of Cell	Name	Location	Function
A			
B			
C			

6. (a) The diagram represents part of a young root shown in longitudinal section. Name the regions of the root labelled A, B, C and D (Fig. 17.3(a)).

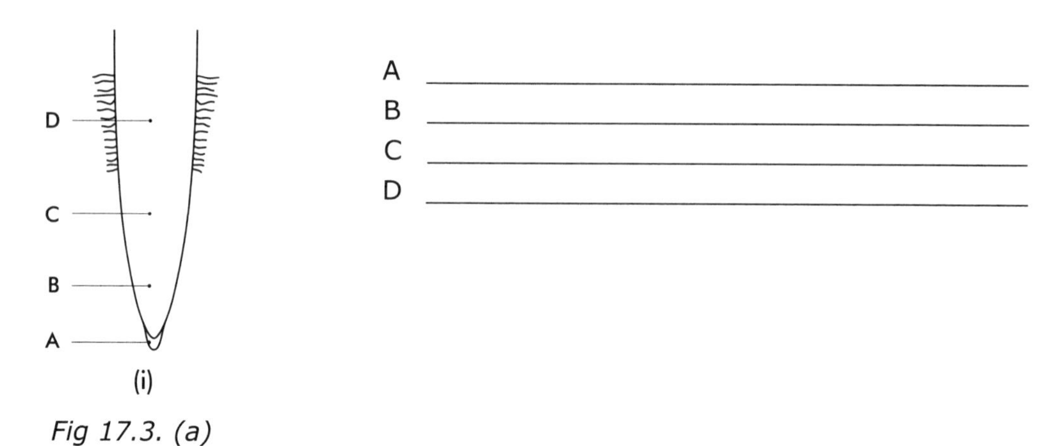

A ______________________________

B ______________________________

C ______________________________

D ______________________________

Fig 17.3. (a)

In which region is the absorption of water carried out? ___________

(b) On the transverse section of a root shown below name the tissues labelled E, F, G, H and I (Fig. 17.3(b)).

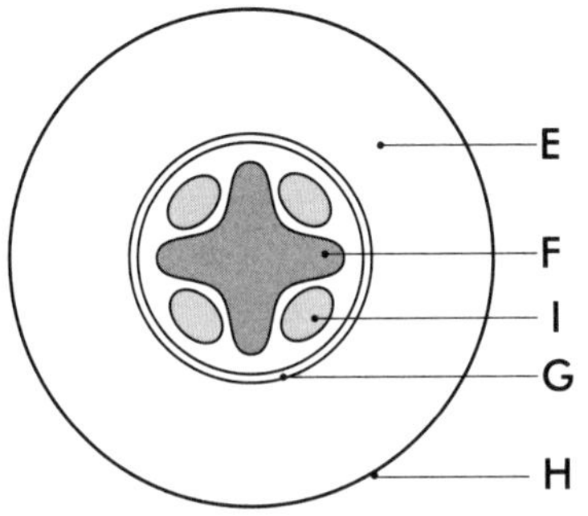

E _______________
F _______________
G _______________
H _______________
I _______________

(ii)

Fig 17.3. (b)

Use letters to indicate which of the tissues shown can be classified as:

(i) Dermal _______________

(ii) Ground _______________

(iii) Vascular _______________

SECTION B

1. (a) Name a dicotyledonous (dicot) plant. _______________

(b) Describe the steps you would take to prepare a transverse section of a dicot stem for examination under the microscope.

2. Answer the following questions concerning the preparation and examination of a transverse section under the microscope:

(a) Why do you use a coverslip? _______________

(b) Which objective lens should you use, initially, to find some cells on the slide?

(c) Why should you not allow your fingers to touch the microscope lens?

____

(d) Describe any safety precautions you might take when preparing the transverse section.

____

____

SECTION C

1. The diagram shows the external appearance of a herbaceous, dicotyledonous plant (Fig. 17.4). This plant is a biennial and has an inflorescence.

(a) Name a dicotyledonous (dicot) plant.
(b) Explain each of the underlined terms.
(c) List three ways by which a dicot can be distinguished from a monocot.
(d) Name the parts labelled A, B, C and D in the diagram and give the function of the part labelled D.

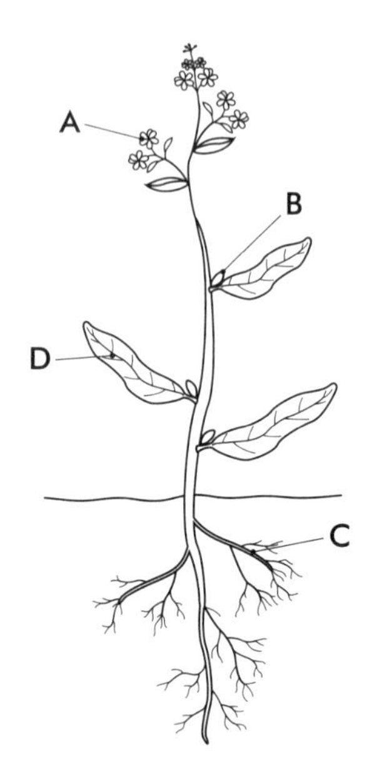

Fig. 17.4

2. Below are four pairs of statements about monocotyledons and dicotyledons. Make a table like the one shown below and sort them into groups.

Monocotyledons	**Dicotyledons**

(a) The leaves have parallel veins.
The leaves have netted veins.
(b) They have two cotyledons.
They have one cotyledon.
(c) The flower parts are in multiples of three.
The flower parts are in multiples of four and five.
(d) They are mainly herbaceous.
There are both herbaceous and woody types.
(e) Their vascular bundles are scattered around the stem.
Their vascular bundles are arranged in an orderly ring in the stem.

3. (a) Draw a large labelled outline diagram of a transverse section through a dicotyledonous stem. Indicate on the diagram the tissue in which water travels to the leaves and the tissue in which soluble foods pass down to the root.

(b) Give two differences between the section you have drawn and a transverse section through the root of the same plant.

(c) Name and draw one example each of the water-transporting and the food-transporting cell types.

(d) Imagine the section of the stem you have drawn was taken from a plant that had its roots placed in water containing a blue dye for 12 hours. Mark in on your diagram the areas that would be blue.

4. (a) 'Xylem is both a vascular tissue and a support tissue.' Explain what is meant by this statement.

(b) Describe how xylem is adapted for its functions.

(c) Draw a diagram of a xylem vessel in longitudinal section. Label the following parts on your diagram: vessel wall, pits, bands of lignin.

(d) In what ways does a xylem vessel differ from a xylem tracheid?

chapter 18 *Transport in the Flowering Plant*

SECTION A

1. Make sentences, in the spaces below, by matching the following phrases. Begin each sentence with a phrase on the left and finish with a phrase on the right.

The upper surface of leaves is covered	by stomata
Evaporation from the surface of the leaves	by the guard cells
The stomata are surrounded	by a waxy cuticle
Transpiration from the stomata can be reduced	is called transpiration
The epidermis is pierced	by the size of the stomata

(a) ______________________________

(b) ______________________________

(c) ______________________________

(d) ______________________________

(e) ______________________________

2. Give a reason for each of the following:

(a) Part of the root of a plant is covered with fine root hairs. ______________________________

____

(b) Root hairs do not possess a cuticle. ______________________________

(c) Stomata usually open in the daylight and close at night. ______________________________

____

3. Answer each of the following:

(a) Give the name of the plant tissue in which water is transported. ______________

(b) The loss of water vapour from the surface of leaves is called ______________

(c) Name the strengthening material, other than cellulose, present in the walls of xylem vessels.

____

(d) State two factors which control transpiration.

(i) ______________________________

(ii) ______________________________

4. The diagram shows part of a section through the leaf of a flowering plant (Fig. 18.1).

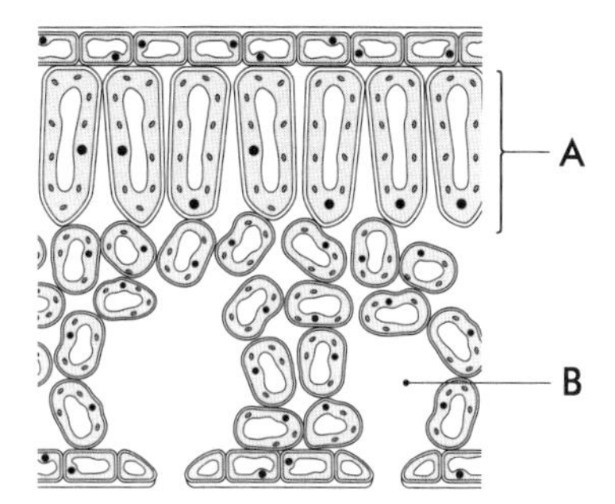

Fig. 18.1

(a) Name the parts A and B.

A ____________________________

B ____________________________

(b) Not all of the water that enters the leaf through the xylem is transpired through the stomata of the leaf. Explain what happens to the water not transpired through the stomata.

____

____

____

5. Briefly state how each of the following are transported in the plant.

Minerals. ____

Carbon dioxide. ____

Water. ____

Glucose. ____

Oxygen. ____

6. (a) Why do plants store food?

____

(b) Where and in what form do the following plants store food?

A carrot plant. ____

A potato plant. ____

A sunflower plant. ____

A leek plant. ____

Sugar beet. ____

(c) How could you demonstrate the presence of the food stores you mention above?

____

____

SECTION B

7. Four leaves from the same plant were weighed and treated as follows:

Leaf 1	Coated with petroleum jelly (Vaseline®) on both surfaces.
Leaf 2	Coated with petroleum jelly on the lower surface only.
Leaf 3	Coated with petroleum jelly on the upper surface only.
Leaf 4	Not coated at all.

The leaves were then weighed, left in bright airy conditions for one hour, and weighed again. State giving reasons:

(a) Which leaf would show the greatest percentage loss in weight? ______________

____

(b) Which leaf would show the smallest percentage loss in weight? ______________

____

(c) Name the part of the leaf which allows it to lose weight? ______________

____

8. An experiment to measure the water loss from and the water uptake by a flowering plant was set up as shown below (Fig. 18.2). The apparatus was weighed at the start of the experiment and again after 24 hours. The results obtained are shown in the table below.

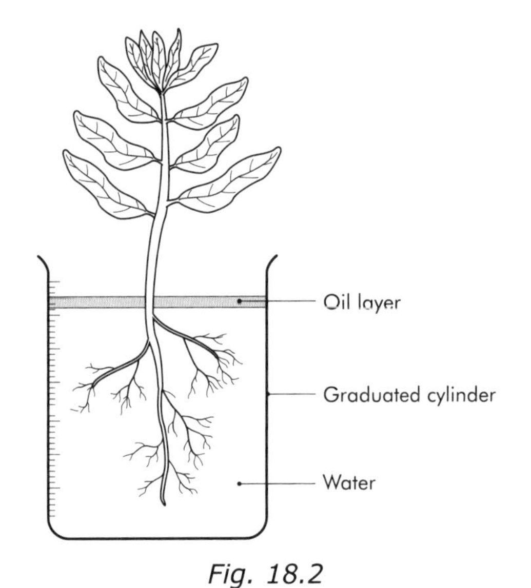

Fig. 18.2

	Mass of apparatus with the plant (g)	**Volume of water in the cylinder (cm^3)**
At the start	210	100
24 hours later	200	86

(a) Calculate the volume of water which has been absorbed by the roots of the plant during the 24-hour period.

____

(b) Calculate the loss in mass due to water loss from the plant during the 24-hour period.

____

(c) What was the purpose of the layer of oil? ______________________

____

(d) Why is the amount of water lost by the plant not the same as the amount of water absorbed by the plant?

____

____

(e) Under certain conditions, a plant growing in a natural environment might lose a lot more water than it absorbs by the roots. Describe the effect this would have on the plant.

____

(f) What control could you have used in this experiment? ______________________

____

SECTION C

1. (a) List three ways in which water is important to plants.

(b) Explain the term transpiration. Why is transpiration needed by plants?

(c) The graph shows the rate of transpiration during a single day (24 hours). Use the graph to give the time during which transpiration was highest. Suggest a reason why transpiration would be highest at this time (Fig. 18.3).

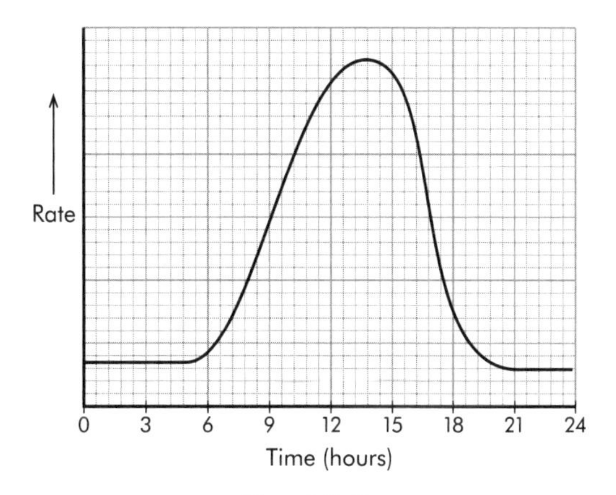

Fig. 18.3

(d) The rate at which transpiration occurs is not unlike the rate at which clothes dry on a washing line. What conditions are needed to dry clothes quickly? In what conditions do clothes dry slowly? List the environmental conditions which you think encourage transpiration to take place.

2. (a) Plants store surplus sugars as starch. They also store lipids as oils. Why do plants store food?

(b) List three structures in which plants store food.

(c) Draw a labelled diagram to show one of the structures used to store food named in (b).

(d) Name a chemical you could use to test for (i) lipids and (ii) starch, stored in a plant.

3. Give an account of the absorption and upward movement of water in the flowering plant.

4. The graph below shows the rate of water uptake and the rate of transpiration by a sunflower plant on a bright, warm sunny day (Fig. 18.4).

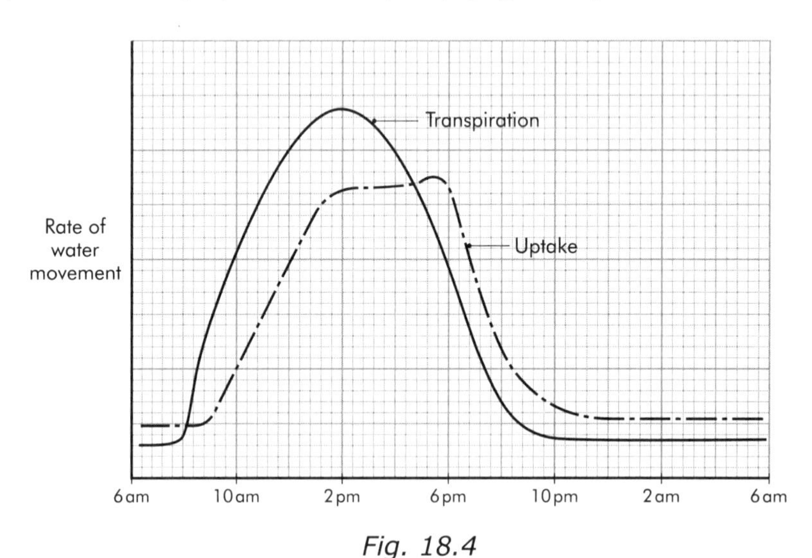

Fig. 18.4

(a) Use the graph to find the time at which each of the following occurs:

(i) The optimum rate of transpiration.

(ii) The rate of water uptake equals the rate of transpiration.

(b) Compare the rate of water uptake with the rate of transpiration from (i) 6 am to 6 pm, and (ii) from 6 pm to 6 am.

(c) Give a reason why (i) the rate of water uptake at 2 pm is lower than the rate of transpiration, and (ii) the rate of water uptake reduces after 6 pm.

(d) Name the pores on the leaf surface through which transpiration occurs and give one other function of these pores in the plant.

chapter 19 *Transport in Humans*

SECTION A

1. (a) In a closed blood system the blood is found in ____________________

(b) Single-celled organisms, such as bacteria, transport materials by means of

____

(c) The pulmonary circulatory system pumps blood from the heart to the ____________________ and back to the heart.

(d) The type of blood vessel that allows materials to pass between the blood and the body cells is a

____

(e) Which type of blood vessel possesses valves? ____________________

2. (a) Name two materials that blood transports to cells.

____________________ ____________________

(b) Name two materials that blood transports away from cells.

____________________ ____________________

(c) What are the three main parts of a closed circulatory system?

(i) ____________________ (ii) ____________________ (iii) ____________________

(d) State one similarity and one difference between an open and a closed circulatory system.

Similarity ____________________

Difference ____________________

(e) Some animals do not have a circulatory system. Name an animal that lacks a circulatory system.

____

(f) Explain how such animals transport materials.

____

3. Which type of blood vessel carries blood

(a) at the highest pressure? ____________________

(b) at the lowest pressure? ____________________

(c) at the slowest speed? ____________________

(d) in pulses? ____________________

4. The diagram shows a transverse section (TS) of an artery (Fig. 19.1).

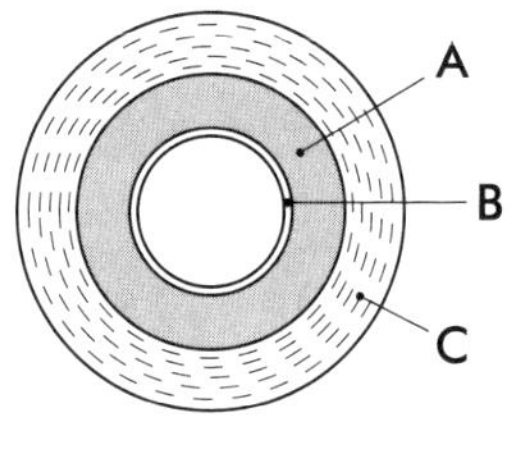

(a) Name the parts labelled A, B and C.

A ______________________

B ______________________

C ______________________

Fig. 19.1

(b) List three ways in which the structure of the artery shown differs from that of a vein.

(i) ______________________

(ii) ______________________

(iii) ______________________

5. In the space below draw a labelled diagram to show the structure of a blood capillary.

What is the main function of blood capillaries? ______________________

____

State three features of capillaries that adapt them to this function.

1. ______________________
2. ______________________
3. ______________________

6. The diagram shows a section through the human heart viewed from the front (Fig. 19.2).

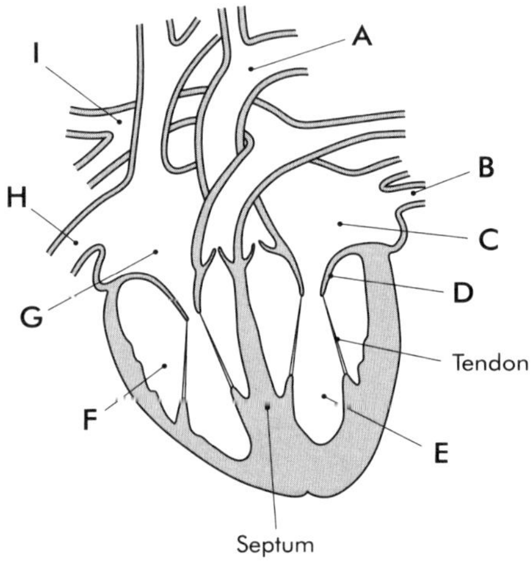

(a) Name the parts of the heart labelled A–E.

A ______________________

B ______________________

C ______________________

D ______________________

E ______________________

(b) Give the appropriate letter to indicate the location of:

(i) the right atrium. ______________________

(ii) the vena cava. ______________________

Fig. 19.2

(c) Which of the following sequences of letters describes the path taken by blood passing through the heart?

IFGHBCDA	☐	AECBHGFI	☐
HGFIBCEA	☐	BCEAIFGH	☐

(d) Why is the wall of chamber C thinner than the wall of chamber E?

____

7. Distinguish between the following pairs:

(a) Open and closed blood circulatory systems. ______________________________

____

(b) Single and double circulation. ______________________________

____

(c) Atria and ventricles. ______________________________

____

(d) Systole and diastole. ______________________________

____

(e) Sino-artial node and atrio-ventricular node. ______________________________

____

(f) Pulmonary vein and pulmonary artery. ______________________________

____

SECTION B

1. In order to learn more about the internal structure of the mammalian heart, a dissection of a sheep's heart can be performed.

(a) Name three pieces of dissecting equipment.

(i) ____________ (ii) ____________ (iii) ____________

(b) Describe how you position the heart for dissection. ______________________________

____

(c) State how you would distinguish the left-hand side from the right-hand side of the heart you are about to dissect.

____

____

(d) Describe in steps how you would dissect a heart to show the main blood vessels and the contents of the heart.

(e) What are flag labels?

Draw an example of a flag label.

(f) Name the arteries through which the heart muscle receives its own blood supply.

(g) Are these arteries visible in the dissected heart?

(h) Blockage of these blood vessels can cause a heart attack. Name a substance that might block these arteries.

State one aspect of our modern lifestyle or diet that doctors believe to be a factor in causing such blockages.

2. The pulse rate tells the number of times the heart beats per minute. It is also known as the heart rate.

(a) What is the pulse?

(b) Describe how to (i) find your pulse and (ii) how to determine your pulse rate.

(i)

(ii) ____

____

____

____

(c) Describe in steps how you would show the effect of exercise on your pulse rate.

____

____

____

____

____

____

____

____

(d) Outline the effect of exercise on your pulse rate. ______________________

____

3. The heart rate of an athlete was recorded before, during and after a race for a total time of 110 minutes. The results are shown in the table below.

Time (min)	0	10	20	30	40	50	60	70	80	90	100	110
Heart rate (beats/min)	62	62	62	63	64	85	107	110	102	90	72	65

(a) Plot a line graph of these results on the grid provided. Put time on the horizontal axis.

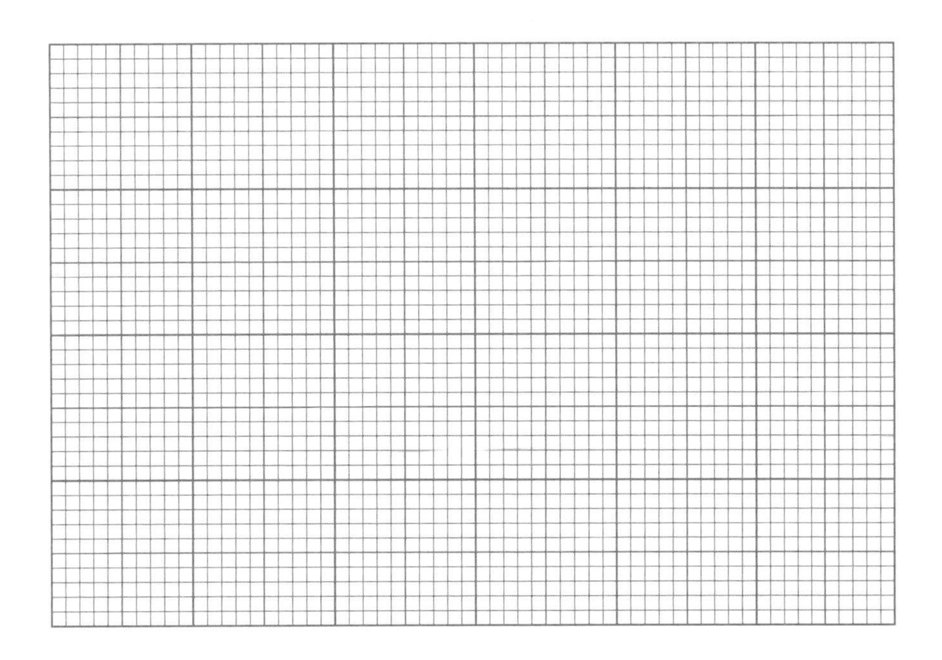

(b) What is the average heart rate at rest? ______

(c) After how many minutes did the athlete begin the race? Give a reason for your answer.

______

(d) After how many minutes did the athlete stop running? Give a reason for your answer.

______

(e) Suggest a reason why her heart rate increased just before she began running.

______

(f) Give two reasons why heart rate must increase during exercise.

(i) ______

(ii) ______

SECTION C

1. (a) Draw a large diagram of the internal structure of the human heart. Include the major blood vessels entering and leaving the heart. Label the following 10 parts: left atrium, vena cava, pulmonary artery, pulmonary vein, semi-lunar valves, tricuspid valve, aorta, pacemaker, septum, right ventricle.

(b) Use arrows to indicate the flow of blood through the heart.

(c) Which of the labelled blood vessels carry oxygenated blood?

2. (a) What is the pacemaker?

(b) What is the function of the pacemaker?

(c) Outline how the pacemaker works.

3. An electrocardiogram (ECG) is a record of the electrical changes that take place as a heart beats. Two electrocardiograms are shown below (Fig. 19.3). The person with the normal ECG had a heart rate of 64 beats per minute. The abnormal ECG was that of a person with a heart condition in which a reduction in the oxygen supply to the heart has weakened the heart muscles. The heart rate of this person was 136 beats per minute.

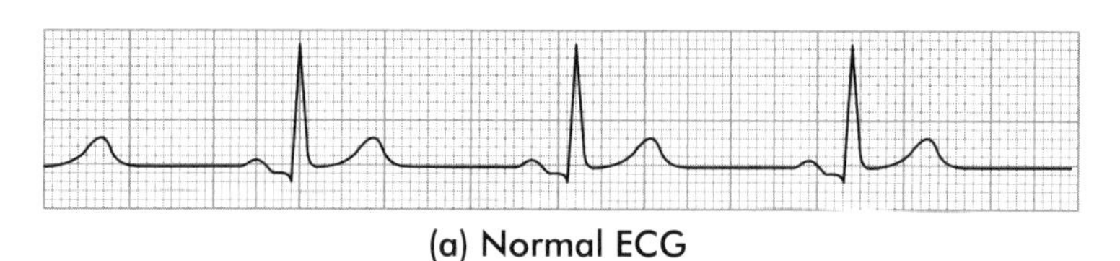

(a) Normal ECG

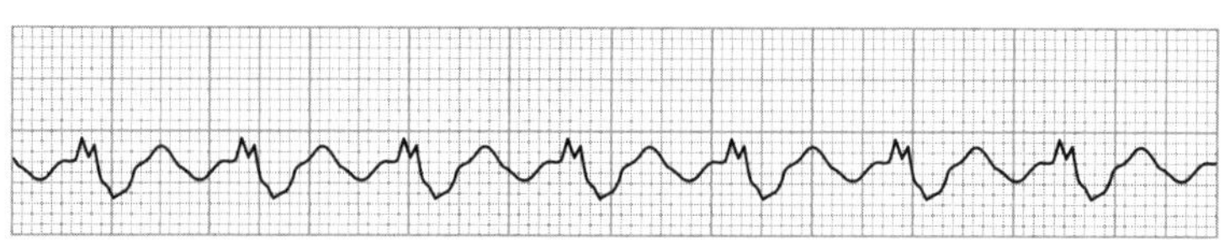

(b) Abnormal ECG

Fig. 19.3

(a) Suggest why the person with the heart condition had a much higher heart rate than the person with the normal heart.
(b) Name three things that can cause heart disease.
(c) Outline how we can help prevent heart disease.

4. (a) What is blood pressure?
(b) Why is blood pressure needed?
(c) The graph (Fig. 19.4) shows the change in pressure in the ventricles during a heartbeat.

(i) What is the maximum pressure in the ventricles?
(ii) What allows the ventricles to produce this high pressure?
(iii) At what time is the pressure rising most rapidly?

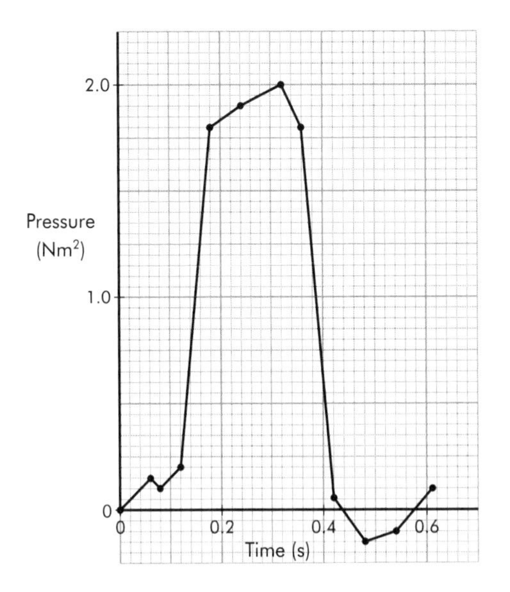

Fig. 19.4

5. The diagram represents the mammalian circulatory system.
(a) Copy the diagram (Fig. 19.5) into your exercise book.
(b) Name the blood vessels labelled A–G (Fig. 19.5).
(c) Label the pulmonary, systemic and portal circulatory systems.
(d) Colour in all those chambers and blood vessels carrying (i) oxygenated blood using a red pen or pencil, and (ii) deoxygenated blood using a blue pen or pencil.
(e) Comment on the validity of the following statement: 'all arteries carry oxygenated blood'.

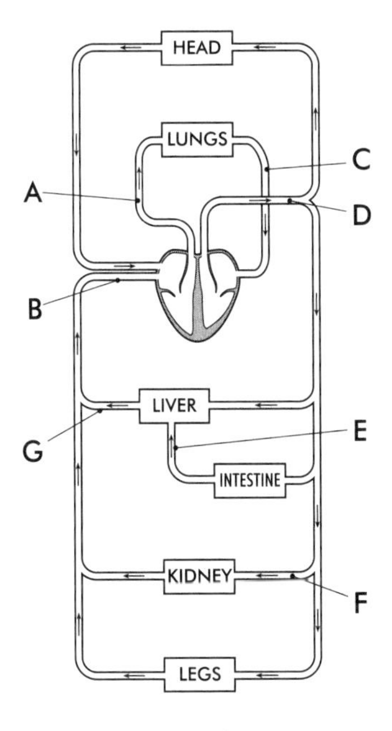

Fig. 19.5

6. Describe the effects of each of the following on the circulatory system:
(a) smoking, (b) diet and (c) exercise.

7. (a) What is lymph?

(b) What is the lymph system?

(c) State three functions of the lymph system.

(d) The diagram (Fig. 19.6) shows the relationship between lymph vessels (lymphatics), body cells and blood capillaries.

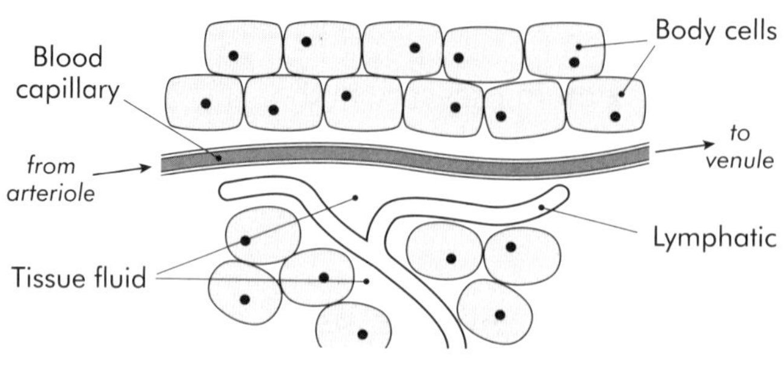

Fig. 19.6

(i) Copy Fig. 19.6 into your exercise book and indicate on the diagram, by means of arrows, the movement of fluid between the blood, the cells and the lymph.

(ii) Mark on the diagram with the letter X a place where you would expect to find the highest blood pressure; and with the letter Y a place where you would find the lowest blood pressure.

(iii) What eventually happens to the fluid that passes into the lymph vessels?

8. (a) Explain the terms systole and diastole.

(b) Give the location of the S-A and A-V nodes in the heart.

(c) Describe the function of the S-A node.

(d) Outline the stages of the human cardiac cycle (heartbeat).

chapter 20 *Blood*

SECTION A

1. Select the figures from column B that most closely approximate to the statements in column A, and insert them in the spaces provided. The figures apply to humans.

	Column A	Column B
(a) ______	Number of white blood cells per mm^3 blood	70
(b) ______	Number of red blood cells per mm^3 blood	4
(c) ______	Adult heart rate per minute at rest	0
(d) ______	Number of chambers in the heart	8,000
(e) ______	Number of nuclei in red blood cells	1
		5,000,000

2. Complete the following sentences:

(a) The gas transported mainly as bicarbonate ions is ______

(b) ______ is the gas mainly transported in the red blood cells.

(c) The platelets function in ______

(d) The pigment found in the red blood cells is called ______

(e) White blood cells are formed in ______

3. Complete the following table:

Substance transported by the blood	From	To
Oxygen	Alveoli	
Wastes, e.g. urea		
Digested foodstuffs, e.g. glucose		Body cells
Hormones, e.g. insulin		

4. (a) Which antigen is present in a person of blood group B? ______

(b) Name the pigment in red blood cells. ______

(c) State the function of the platelets. ______

(d) Give one location in the body where white blood cells are formed. ______

(e) Name two substances, other than wastes and gases, which are transported by the blood:

(i) ______

(ii) ______

SECTION C

1. Describe clearly the difference between red blood cells, white blood cells and platelets under the following headings: (i) structure, (ii) function and (iii) site of formation.

2. The diagram shows the blood cells (Fig. 20.1).

(a) Copy and label a red blood cell and a white blood cell as seen in the diagram.

(b) Name three ways in which red blood cells are different from white blood cells.

(c) State two features of red blood cells that adapt them for their function in carrying oxygen.

(d) What are the functions of the white blood cells?

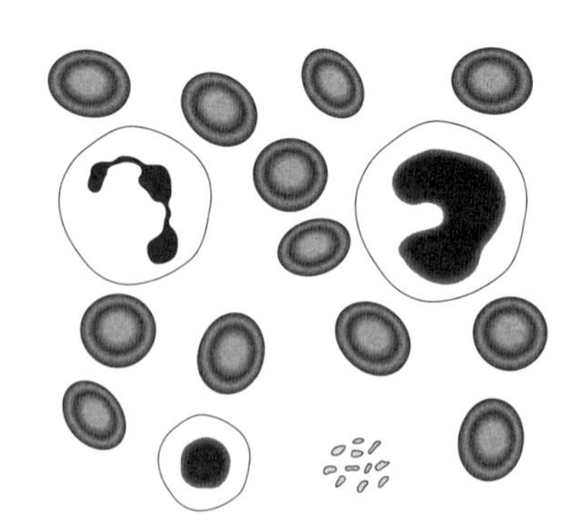

Fig. 20.1

3. The table shows the blood composition of four people.

	Red blood cells (per mm^3)	White blood cells (per mm^3)	Platelets (per mm^3)
Mark	7,750,000	550	250,000
Sarah	2,200,000	4,000	600
Donal	5,200,000	6,000	260,000
Matt	5,000,000	8,000	250,000

(a) One person is suffering from anaemia (lack of iron in the diet). Who do you think it is? Give a reason for your choice.

(b) Give one symptom of anaemia.

(c) Which person has blood that will take a long time to clot? Explain your choice.

(d) At high altitude, e.g. the Himalayas, there is less oxygen in the air than at sea level and more red blood cells are needed to carry oxygen efficiently. One of the people in the table lives at high altitude. Which person do you think it is? Explain your choice.

4. Explain the reason for each of the following.

(a) When you are suffering from certain illnesses, the number of white blood cells in your bloodstream is greater than at other times.

(b) The importance of determining a person's blood group before giving a blood transfusion.

(c) During pregnancy, a woman is often given iron tablets.

(d) When viewed under a light microscope, the middle part of a red blood cell appears paler than the outer part.

chapter 21 *Animal Nutrition*

SECTION A

1. Match the terms in column 1 with the explanations in column 2.

Column 1	**Column 2**
Autotrophs	Organisms which feed on plants and animals.
Carnivores	Organisms which feed on food made by some other organism.
Heterotrophs	Organisms which feed on other living things, often causing harm.
Herbivores	Organisms which feed on dead plant and animal material.
Omnivores	Organisms like green plants which make their own food.
Parasites	Organisms which feed on plants.
Saprophytes	Organisms which feed on animals.

2. Give one example each of the feeding types listed below:

Feeding type	**Named example**
Autotroph	
Carnivore	
Heterotroph	
Herbivore	
Omnivore	
Parasite	
Saprophyte	

3. The diagram (Fig. 21.1) shows part of the human digestive system. Label the parts A–F.

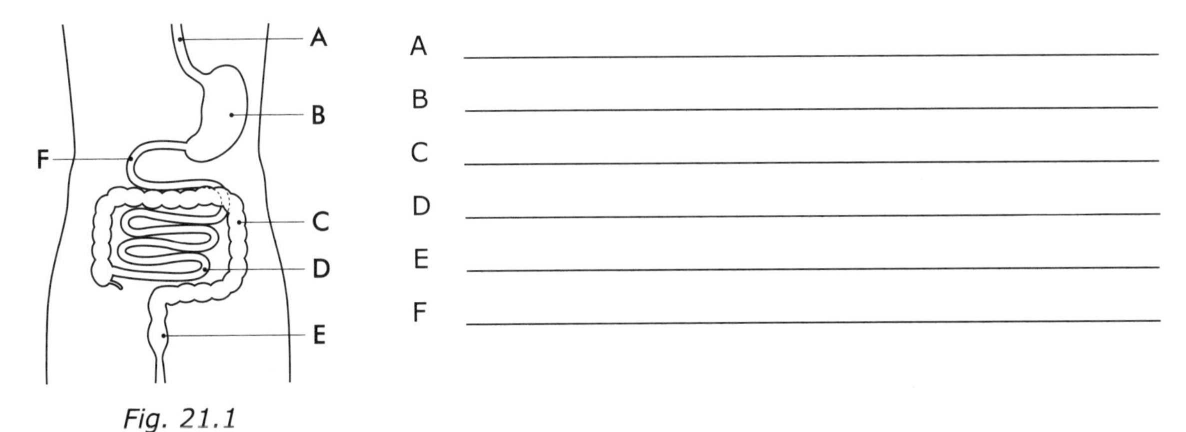

Fig. 21.1

Mark X on the diagram to show the location of the liver.

Name a part of the digestive system that is not shown in the diagram.________________

4. Mechanical digestion is the physical breakdown of food into smaller pieces. Chemical digestion is the breakdown of insoluble food into soluble form by means of enzymes.

(a) Why is mechanical digestion necessary?________________________________

____

(b) Sort the following list of statements into the correct column in the table below:

action of peristalsis   action of lipase   action of saliva
action of teeth   action of bile   action of stomach acid

Mechanical digestion	Chemical digestion

5. (a) State the number of incisor teeth in the full adult set.______________________

(b) Which type of tooth is not present in the milk set?________________________

(c) Which part of a tooth contains blood vessels?____________________________

(d) Give the dental formula of an adult human with a full set of permanent teeth.

____

(e) What is the function of the canine teeth?_______________________________

6. For each of the functions listed below, name a part of the digestive system to match.

Function	Named part
Produces acid	
Makes bile	
Connects the throat to the stomach	
Has no digestive use in humans	
Where most absorption occurs	
Stores bile	
Where a lot of water is absorbed	
Controls the movement of food out of the stomach	
Stores faeces	
Where food moves by peristalsis	
Where most digestion occurs	
Symbiotic bacteria are found here	

7. (a) Name the tube that carries food from the throat to the stomach. ______________

(b) Name two enzymes produced by the pancreas. ______________

(c) State the function of bile in digestion. ______________

(d) Name a part of the digestive system with a pH > 7. ______________

(e) Give two functions of the liver other than the production of bile.

____

8. What are the benefits of each of the following dietary recommendations?

(a) Increase the amount of fibre eaten per day. ______________

____

(b) Eat less sugar. ______________

____

(c) Increase the amount of fruit and vegetables. ______________

____

(d) Reduce the amount of salt. ______________

____

(e) Reduce the amount of fat, particularly animal fats. ______________

____

SECTION B

1. A student carried out the experiment described below.

(i) Three test tubes were set up as follows:

Tube 1 1% starch solution with 1 cm^3 amylase solution
Tube 2 1% starch solution and 1 cm^3 boiled amylase solution
Tube 3 1% starch solution with 1 cm^3 amylase solution plus 0.5 cm^3 acid

(ii) All three test tubes were placed in a water bath at 37°C for 20 minutes.
(iii) After this time samples were taken from each test tube and tested with iodine solution and Benedict's solution.
(iv) A table of results was completed.

Answer the following:

(a) In the space below draw a labelled diagram of the apparatus set up as described above.

(b) Why was the water bath kept at 37°C?____________________

(c) Complete the table by filling in the results you would expect the student to get in each test tube.

	Colour in Tube 1 after 20 min	**Colour in Tube 2 after 20 min**	**Colour in Tube 3 after 20 min**
Tested with iodine solution			
Tested with Benedict's solution			
Conclusion			

(d) What safety precautions should be taken when doing the Benedict's test?

____

(e) Name two places in the human body where amylase is produced.

____

(f) Where in the human digestive system would conditions like those in Tube C be found?

____

(g) Which test tube represents the control in this experiment?____________________

2. One hundred students took part in an investigation to measure the pH of their saliva and compare it with the number of their teeth which had been filled. The table shows the results of the investigation.

pH of saliva	No. of students in each pH range	Average no. of fillings per student in each pH range
7.4 – 7.2	22	3
7.2 – 7.0	58	5
7.0 – 6.8	20	7

(a) What percentage of students have saliva in the 7.2–7.0 range? ______________

(b) Describe how you would measure the pH of the student's saliva.

____

(c) How might eating an orange affect the pH readings? ______________________

____

(d) What hypothesis is being tested in this investigation?

____

(e) Do the results explain the hypothesis? Give a reason for your answer.

____

____

(f) What conclusions can you draw from the results? ______________________

____

(g) Mention one way in which the investigation could be improved.

____

SECTION C

1. (a) Explain the term digestion.

(b) Outline the need for digestion and a digestive system.

(c) Distinguish between mechanical and chemical digestion. Give two examples of mechanical digestion.

2. (a) Give three reasons why animals need food for survival.

(b) Spiders and certain insects are carnivores.

(i) What is a carnivore?

(ii) Predict what kind of digestive enzymes these insects and spiders would produce. Give a reason for your answer.

3. (a) What is meant by peristalsis?
(b) Where does peristalsis occur?
(c) What is the function of peristalsis in the digestive system?

4. (a) Starting at the mouth, list, in the correct order, the pathway of a piece of apple skin until it reaches the large intestine.
(b) What name is given to material, such as the skins of fruit and the stalks of vegetables, which is not fully broken down in the digestive system? Why is it important to include such material in our diet?

5. (a) Where is the gall bladder found?
(b) Sometimes a person's gall bladder has to be surgically removed. Suggest what effect this would have on the person's digestion and state how the person might have to alter their eating habits.

6. Describe (a) the location and (b) the function of each of the following in the human alimentary canal:

incisors villi pyloric sphincter muscle symbiotic bacteria gastric glands

7. A person eats a prawn curry, some rice and chips and drinks a can of fizzy lemon drink.
(a) What are the main nutrient types (i.e. carbohydrate, fats, protein) in (i) prawns, (ii) rice, (iii) cooking oil and (iv) the fizzy drink?
(b) In what form are these nutrients when they are finally absorbed into the bloodstream from the ileum and to what use does the body put them?
(c) Name an enzyme involved in the digestion of protein and state where it is produced.
(d) Do you consider the meal described as a healthy meal? Give reasons for your answer.

chapter 22 *Gas Exchange in Organisms and the Human Breathing System*

SECTION A

1. (a) Explain the term gas exchange. ______________________________

(b) Give three reasons why gas exchange is necessary in living organisms.

____

(c) List three features of an efficient gas exchange surface.

____

____

____

2. (a) Name the parts labelled A–D on the diagram of a section through a leaf (Fig. 22.1).

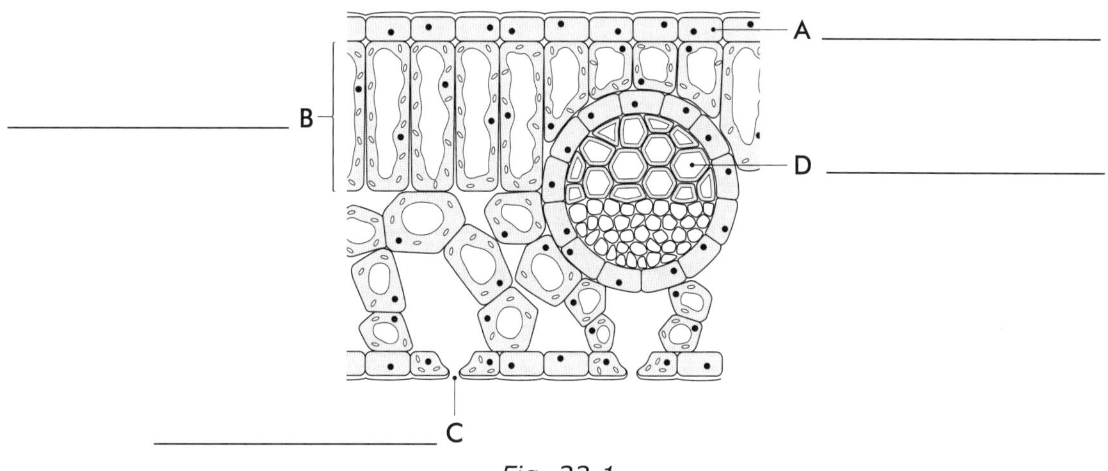

Fig. 22.1

(b) Through which labelled part are gases exchanged? ______________________________

(c) List three ways in which leaves are adapted for rapid exchange of gases.

(i) ______________________________

(ii) ______________________________

(iii) ______________________________

3. (a) Complete the table below to compare gas exchange in humans with that in a flowering plant, such as the buttercup or grass.

Feature	Human being	Buttercup/grass
Gases exchanged		
Body part/organ involved		
One adaptation for rapid gas exchange		
Another adaptation for rapid gas exchange		

(b) The buttercup is not a woody plant. Name the structures found in woody stems that allow gases to be exchanged.

____

4. Label the parts A–H on the diagram of the human breathing system below (Fig. 22.2). Give the function of the parts labelled D, E and H.

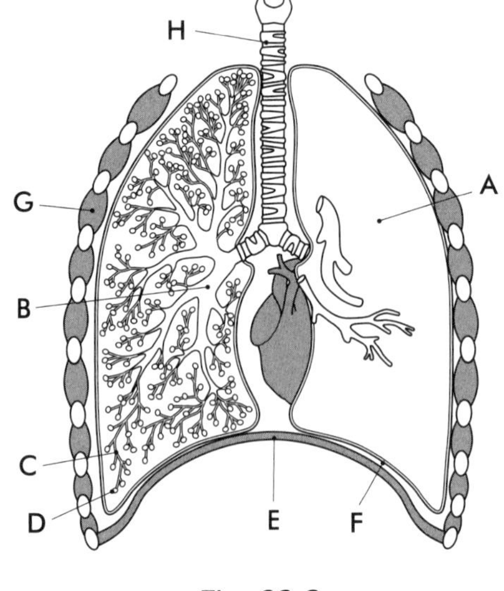

Fig. 22.2

A ______________________________

B ______________________________

C ______________________________

D ______________________________

Function: ______________________________

E ______________________________

Function: ______________________________

F ______________________________

G ______________________________

H ______________________________

Function: ______________________________

5. (a) Gas exchange in humans occurs at the ______________________________

(b) Give the percentage (%) oxygen in expired air. ______________________________

(c) The blood carries oxygen in the ______________________________

(d) The average rate of breathing of an adult human at rest is ______________________________

(e) Gas exchange between woody stems and the atmosphere occurs through __________

(f) Intercostal muscles function in the process of ______________________________

(g) The bronchioles join the bronchi and the ______________________________

(h) Gas exchange between a leaf and the atmosphere occurs through the

____

6. Complete the following by choosing the most appropriate term and putting a line through the incorrect term.

When you exercise your breathing is faster/slower and deeper/more shallow. The lungs/brain control breathing. When you exercise you make more carbon dioxide/oxygen. The amount of this gas builds up in the blood and the lungs/brain detect(s) the rise. A message is sent to your heart/chest muscles which causes your breathing rate to decrease/increase to lower/raise the oxygen/carbon dioxide levels in your blood.

SECTION B

1. The graph below shows a girl's rate of breathing (Fig. 22.3).

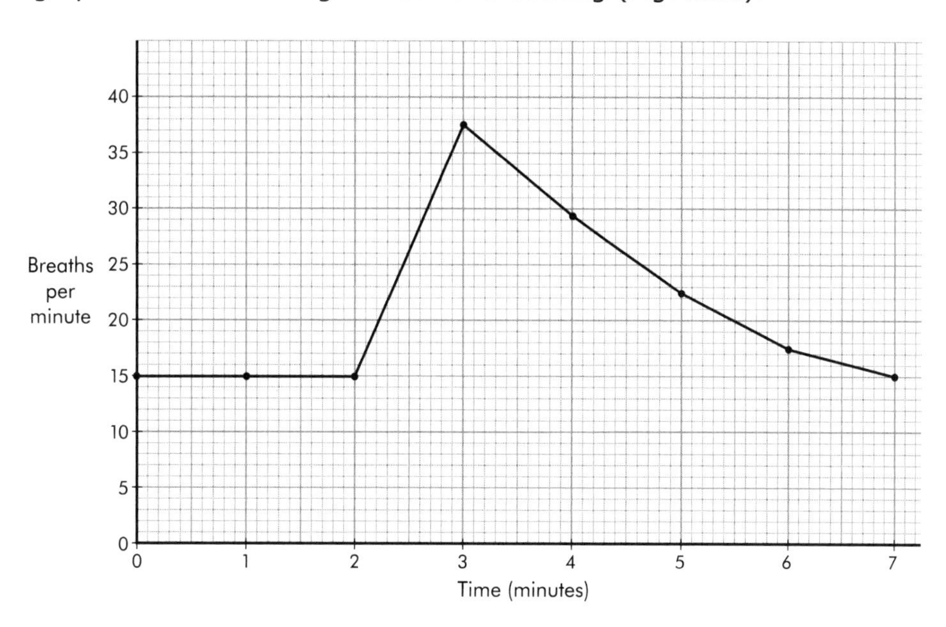

Fig. 22.3

(a) How would you measure the girl's rate of breathing? ______________________

____

____

(b) During which period of time was the girl carrying out strenuous exercise?

____

(c) Use the graph to find the girl's breathing rate (i) at rest and (ii) immediately after exercise.

(i) ______________________ (ii) ______________________

(d) How long did it take for the breathing rate to return to 15 breaths per minute.

____

2. Examine the apparatus shown in the diagram Fig. 22.4. The apparatus was kept at 30 °C. After about 20 minutes the lime water had turned a milky colour.

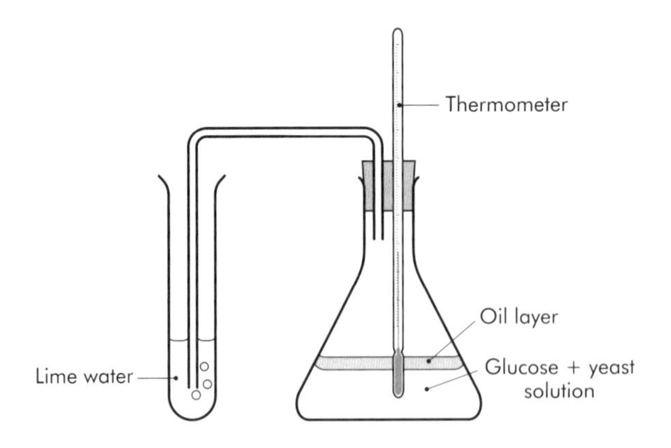

Fig. 22.4

(a) What gas is being produced by the yeast? ______________________

(b) Name the process that produced the gas. ______________________

(c) Why was the temperature kept at a constant 30°C? ______________________

(d) State how you would keep the temperature constant. ______________________

____

(e) Suggest a control for this experiment. ______________________

____

3. The diagram Fig. 22.5 shows a simple smoking machine. Bicarbonate indicator responds to the presence or absence of acid substances. It is a cherry-red colour when neutral, yellow in the presence of an acid substance, and purple if there is less acid around. The apparatus is set up as shown and the suction pump is turned on for five minutes.

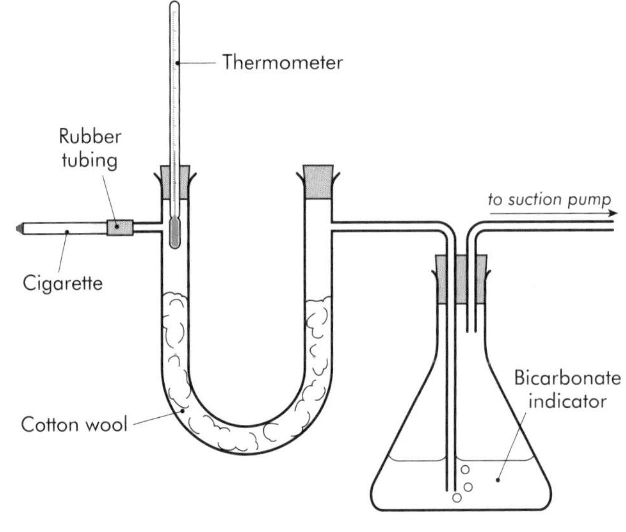

Fig. 22.5

(a) The colour of the bicarbonate indicator changed from cherry red to yellow. Why do you think this happened?

____

(b) The cotton wool becomes a yellow/brown colour. What does this show?

____

____

(c) Describe what you think the smell of the cotton wool would be like after the cigarette has been 'smoked' for five minutes.

____

____

(d) Describe a control for this experiment. ______________________

____

What is the purpose of setting up a control? ______________________

SECTION C

1. (a) During breathing we use the <u>diaphragm</u>, the <u>intercostal muscles</u> and the <u>brain</u>. Describe the breathing mechanism with reference to the words underlined (diagrams not required).

(b) Explain briefly how the exchange of gases occurs in plants.

2. (a) Draw a labelled diagram of one alveolus together with its blood supply.

(b) Indicate on your diagram each of the following: (i) the direction of blood flow; (ii) a region that is high in oxygen; and (iii) a region that is low in oxygen.

(c) Name the method of gas exchange in the alveoli.

(d) State one way in which the blood transports carbon dioxide.

(e) Give two features of the alveoli that help gas exchange.

3. The information below was printed on the side of a packet of cigarettes.

WARNING: SMOKING CAN SERIOUSLY DAMAGE YOUR HEALTH

(a) Name one disease of the respiratory system which could be caused by smoking.

(b) Describe the disease you named in (a) under the headings: possible cause; prevention; and treatment.

(c) The motion for the next school debate has the title 'Smoking won't affect me'. Write a short speech to (i) support and (ii) reject the motion.

4. The graph Fig. 22.6 shows the volume of air breathed in and out during three different activities.

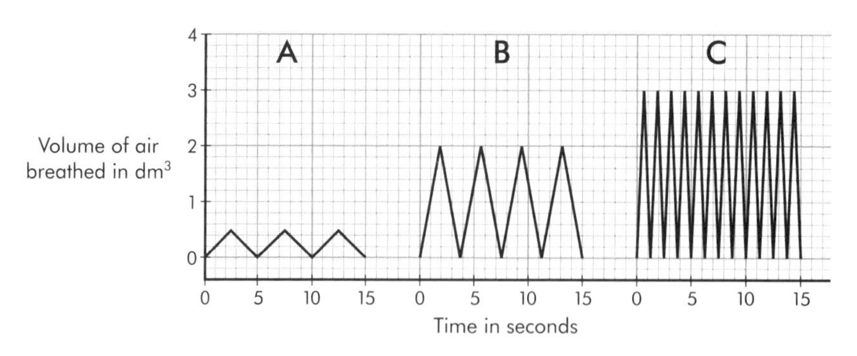

Fig. 22.6

(a) What was the volume of air breathed in and out per breath (i) during activity A, and (ii) during activity C?

(b) What was the rate of breathing (number of breaths per minute) during the period of activity B? Compare this rate of breathing with that during activity C.

(c) What two features on the graph indicate that exercise is taking place?

(d) Which activity A, B or C represents the person (i) cycling hard, and (ii) resting?

(e) List four differences between inspired and expired air.

5. Describe an experiment to determine the relationship between the rate of breathing and exercise level. Comment briefly on the results you would expect to obtain.

chapter 23 *Homeostasis and Excretion*

SECTION A

1. (a) Define the term homeostasis.

(b) Why is homeostasis necessary in multicellular animals?

(c) What are the main variables that are controlled in homeostasis?

(d) Define the term excretion.

(e) What role does excretion play in homeostasis?

2. (a) What is a lenticel?

(b) What role does a lenticel play in excretion?

(c) Why is it necessary to have a lenticel?

(d) What structure in the leaf has the same function as a lenticel?

(e) What waste gas is produced by a plant in (i) the dark and (ii) daylight?

(i)

(ii)

(f) Why is there a different waste gas under different light conditions?

3. In the apparatus shown below (Fig. 23.1) the excretion of carbon dioxide by a plant is demonstrated. Sodium hydroxide will absorb carbon dioxide and lime water will turn 'milky' in the presence of carbon dioxide.

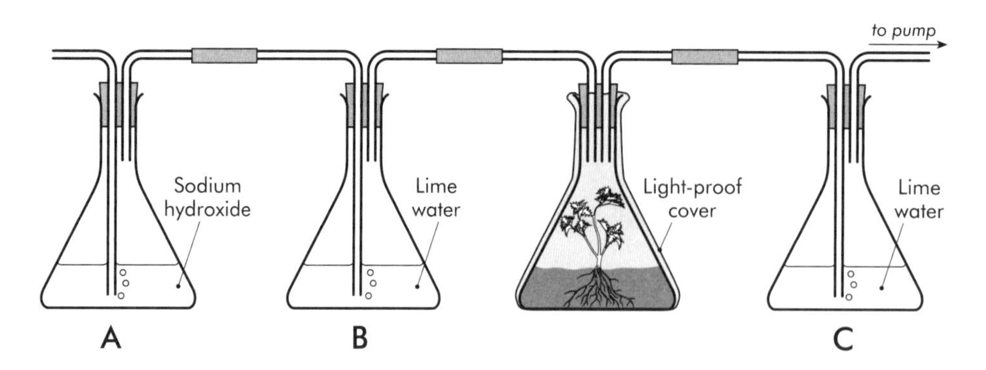

Fig. 23.1

(a) What is function of flask A?____________________

(b) What will happen to the liquid in flask B?____________________

(c) Why will this be the case?____________________

(d) What is the purpose of the lightproof cover over the plant?____________________

(e) What should happen to the liquid in flask C?____________________

(f) What does this tell you about the plant?____________________

(g) What would you do to set up a control for this experiment?____________________

4. (a) What are the three excretory organs in humans and what are their excretory products?

____

(b) What is the role of each of these organs in homeostasis?

(c) Describe what happens in the skin when a person's body temperature falls.

(d) What changes occur when the body temperature rises?

5. (a) Label the following diagram of the urinary system in humans (Fig. 23.2).

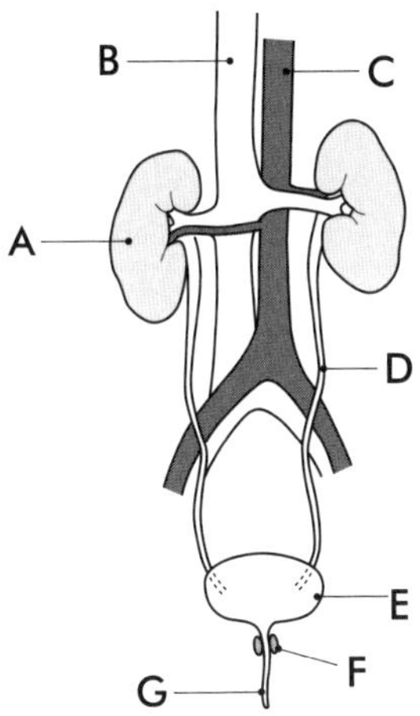

A ____________________

B ____________________

C ____________________

D ____________________

E ____________________

F ____________________

G ____________________

Fig. 23.2

(b) What are the functions of the following?

C ____________________

D ____________________

E ____________________

F ____________________

6. (a) Name the functions of the kidneys. ____________________

(b) What is osmoregulation? ____________________

(c) What is filtration and where does it occur in the kidney?

(d) What is the main function of the medulla in the kidney?

(e) What will the kidney do if there is too much salt in the blood?

(f) What would happen in the kidneys if a person were to drink a large amount of water?

7. The diagram below (Fig. 23.3) shows cells that are found in high numbers on the underside of a leaf.

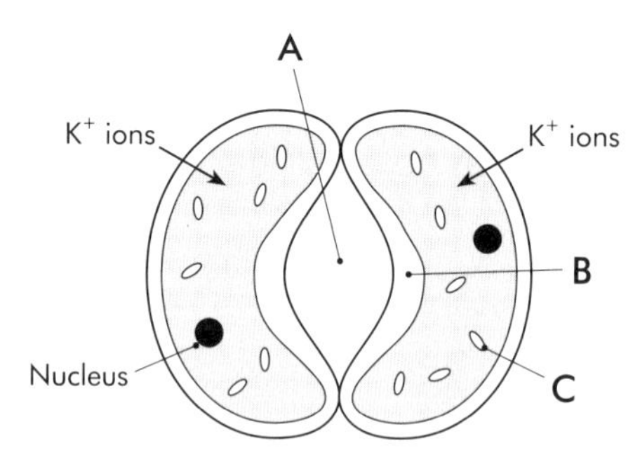

Fig. 23.3

(a) Label the following parts:

A

B

C

(b) In what way are these cells modified?

(c) How is this modification useful to the cell?

(d) It has been suggested that the intake of potassium ions is necessary for the opening of the stoma. By what process would these ions enter the cells?

(e) Why might photosynthesis be necessary to let this process occur?

(f) Why would high levels of potassium ions lead to the influx of water into the cell?

___

(g) What happens at night to the potassium ions and what effect does this have on the cells?

___

___

___

8. (a) What is the similarity between the control of gaseous exchange in a plants and humans?

___

___

___

(b) What part of the brain is involved with this control?___

(c) What is being monitored in the control of the breathing rate?___

(d) Which types of receptors are used to monitor the blood?___

(e) In intensive care facilities in hospital a person is pronounced dead when they have no brain stem activity. Why do you think this is the case?

___

___

9. (a) Label the following diagram of the nephron of the kidney (Fig. 23.4) and give the function of each part.

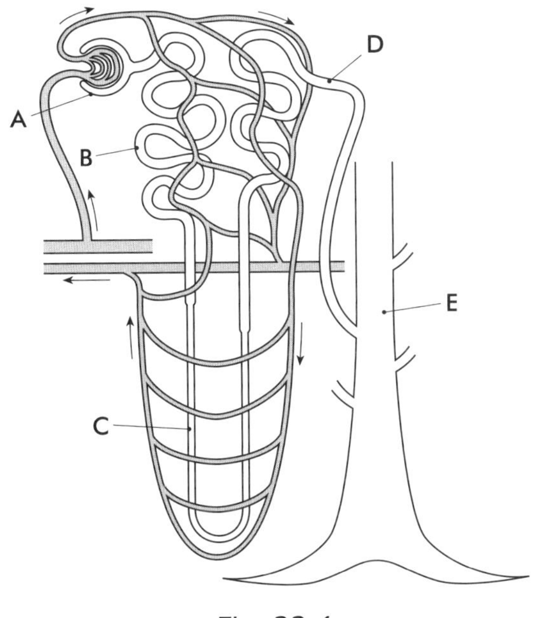

Fig. 23.4

A ______________ ______________________

____

B ______________ ______________________

____

C ______________ ______________________

____

D ______________ ______________________

____

E ______________ ______________________

____

(b) (i) What is pressure filtration? ______________________

____

(ii) How does this occur? ______________________

____

(c) (i) Why are proteins not found in the section marked C? ______________________

____

(ii) Why is glucose not found here either? ______________________

____

(d) Why does the kidney require a lot of energy?

____

(e) What is the function of ADH? ______________________

(f) Under what conditions will the pituitary gland release ADH? ______________________

____

SECTION C

1. (a) What is homeostasis and why is it important for multicellular organisms?
(b) What sort of environmental conditions are controlled in homeostasis?
(c) Compare excretion in plants with that in animals.

2. (a) Describe the role of the skin in temperature regulation in humans.
(b) What is the excretory function of skin?
(c) Dinosaurs, it is believed, were warm blooded, as are mammals, but when the global temperature dropped the dinosaurs died out but the mammals did not. Suggest a characteristic of mammals that may have given them an advantage over dinosaurs. How might this have helped them survive?

3. (a) Draw a simple labelled diagram of the urinary system in humans.
(b) Describe in simple terms how the kidneys function.
(c) Apart from its role in excretion, what is the function of the kidney in homeostasis?
(d) What would be the effect of (i) a protein rich meal, (ii) a hot day and (iii) the intake of salt, on the urine produced by the kidney?

4. The table below gives the concentration of various chemicals in blood plasma and in urine.

	Concentration in plasma (g/ml)	**Concentration in urine (g/ml)**
Water	910.0	950.0
Proteins	75.0	0.0
Glucose	1.0	0.0
Salt	3.7	6.0
Urea	0.3	20.0

(a) Explain the differences between the concentrations of protein and glucose in plasma and urine.
(b) Which of the substances listed above are excreted by the kidneys?
(c) Under what conditions might a person's salt level in the urine increase?
(d) If a person is not producing the hormone insulin then their urine will be found to contain glucose. Why does this happen?

5. (a) Describe the functioning of the nephron in humans.
(b) A gerbil (desert rat) is considered an ideal pet because its cage does not need to be cleaned as often as other rodents because it does not urinate as frequently. It has also been noted that gerbils have particularly long Loops of Henle. Can you relate these two facts?
(c) What is the effect of the hormone ADH on the kidney?
(d) Under what conditions would a person produce (i) increased amounts, and (ii) reduced amounts of ADH?

chapter 24 *Plant Response to Stimuli*

SECTION A

1. (a) Define the term tropism? ______________________________

(b) What is a growth regulator? ______________________________

(c) Name one common growth regulator. ______________________________

(d) List four tropisms and give their effect on a plant.

(i) ____________ ______________________________

(ii) ____________ ______________________________

(iii) ____________ ______________________________

(iv) ____________ ______________________________

(e) How will plants respond to light shining on them from one side?

____

(f) What benefit is this response to a plant? ______________________________

2. (a) What is phototropism? ______________________________

____

(b) What effect does this have on a plant? ______________________________

____

(c) How does this effect differ from geotropism? ______________________________

____

(d) Why do the root and shoot of a plant respond differently to gravity?

____

____

(e) What are the benefits of these responses for a plant?

____

____

3. Some plant growth regulators promote growth and others inhibit growth. Two such growth regulators are involved in leaf fall in trees.

(a) Name these two regulators. ____________ ____________

(b) Which of these regulators is produced early in the year? ____________

(c) What effect does it have on the leaf?

____

(d) Which regulator is produced in the autumn?______________________

(e) What effect does this have on the leaf? ______________________

(f) What is the effect of the artificial growth regulator 2-4-D?

____

(g) What benefit is this in gardening?

____

____

4. Give an example of the effect of each of the following growth regulators and give a use that people have for each.

(a) Ethene.

____

____

____

(b) Gibberellins.

____

____

____

(c) 2-4-D.

____

____

____

5. Give three examples of protective structures found on plants and describe how they are beneficial to the plant.

(a) ______________________

____

(b) ______________________

____

(c) ______________________

____

6. (a) What is a coleoptile? ______________________

(b) Where is auxin produced?______________________

(c) What effect does the auxin have on the cells in the stem of the plant?

____

(d) If you increase the concentration of auxin what will be the effect?

____

(e) How does this affect the plant growth?

____

____

(f) Why is it unlikely that auxins affect roots in plants in the wild?

____

____

(g) What plant growth regulator is responsible for geotropism?______________

(h) What famous scientist was involved in the investigation of phototropism?

____

(i) What did he discover?

____

____

(j) What causes this effect?______________________________

____

7. In an experiment three young ash saplings were treated as follows in early spring (Fig. 24.1):

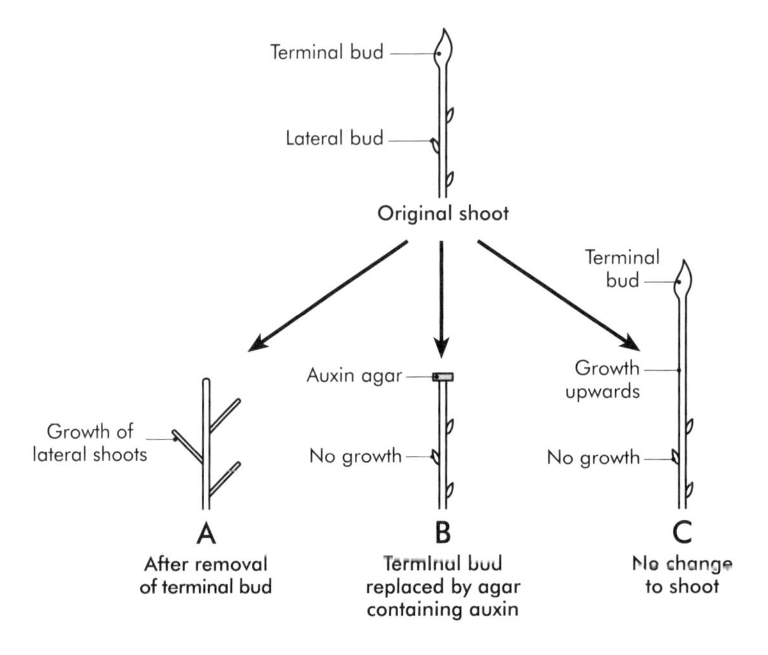

Fig. 24.1

Plant A	Its terminal shoot was removed.
Plant B	Its terminal shoot was removed and a piece of agar containing auxin was placed on top of the cut stem.
Plant C	It was left alone.

After two months the saplings were examined and the following results were found:

Plant A	Lateral buds had grown.
Plant B	No lateral bud growth.
Plant C	Terminal bud growth and no lateral bud growth.

(a) What process was being examined in this experiment?

____

(b) What controls this growth response?

____

(c) Which of the three ash saplings is acting as a control? ______________

(d) What did plant B demonstrate?

____

____

(e) What would be the benefit of this growth effect in an ash plant?

____

____

(f) How is this effect utilised by gardeners when they cut a hedge?

____

____

SECTION B

1. Give an explanation for each of the following steps in an investigation to show the effect of IAA on plant tissue.

(a) Moist filter paper was placed into two petri dishes, A and B.

____

____

(b) About 30 seeds were sprinkled into each petri dish.

____

____

(c) The dishes were put in an incubator set at 30°C for 2–3 days.

____

(d) Each seed was then placed on a glass slide on a piece of graph paper.

(e) The seedlings were placed on fresh filter paper in each of two clean petri dishes as follows:

Dish A had 10 cm^3 of a 10 ppm solution of IAA added.

Dish B had 10 cm^3 of distilled water added.

(f) The seedlings must be kept in the dark.

(g) What was the effect of IAA on the cress seedlings?

SECTION C

1. (a) Describe the different types of tropisms that affect flowering plants and give the benefits of these responses to the plant.

(b) What is the function of plant growth regulators?

(c) Describe the role of plant growth regulators in leaf fall and in fruit ripening.

(d) Give one example of the use humans have made of plant growth regulators and describe how they are used.

2. In an experiment three pots of cress seedlings were placed into three containers as shown in the diagram (Fig. 24.2). After a week the plants were examined.

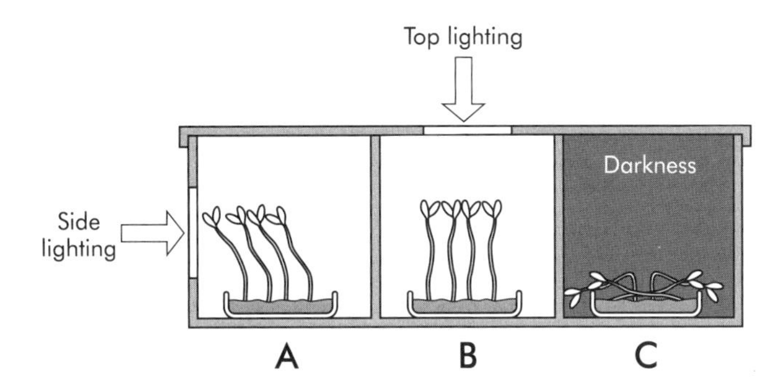

Fig. 24.2

Plants in section A	had elongated in the direction of the opening.
Plants in section B	were slightly longer and growing straight up.
Plants in section C	were yellow and had elongated enormously.

(a) What is being examined in this experiment?

(b) Give a concise explanation for the results in each of the pots?

(c) When gardeners are planting many seeds they usually plant them close together in a seed tray, and a week or so after they have sprouted they are re-potted in a new tray where they are spread out. Using the information from the above experiment can you give a reason for this?

3. A student wished to demonstrate the effect of 2-4-D on a lawn. Explain why she carried out each of the following procedures:

(a) The lawn was divided into two (section A and section B).

(b) A quadrat was thrown at random in each section and the number of dicot plants present was counted after each throw.

(c) A selective weed killer was spread on section A of the lawn.

(d) A month later step (b) was repeated.

In her experiment the student divided her lawn into two 20 m^2 sections. Her quadrat was 0.5 m^2 and she got the following results.

Section	No. of throws taken each time	No. of weeds before treatment	No. of weeds after treatment
A	10	28	5
B	10	32	29

(e) How many weeds were there in section A of the lawn before the treatment?

(f) What was the average number of weeds in each square metre (m^2) of the lawn?

(g) What was the number of weeds in section A after the treatment?

(h) What was the average number of weeds in each square metre of section A after the treatment?

(i) What was the percentage reduction of weeds in section A?

4. (a) Describe the effect of auxins on the shoot tip of plants.

(b) In an experiment oat coleoptiles were treated as follows (Fig. 24.3).

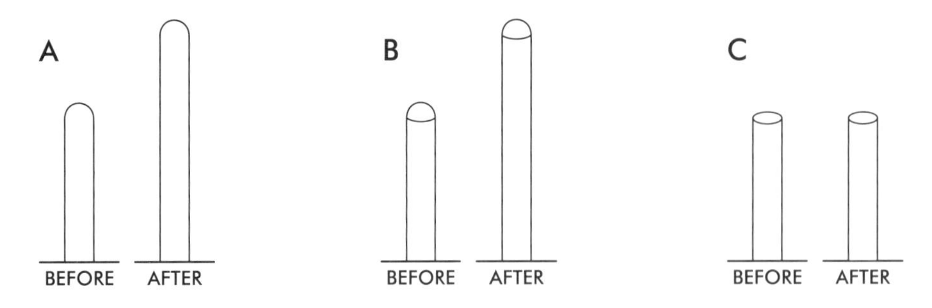

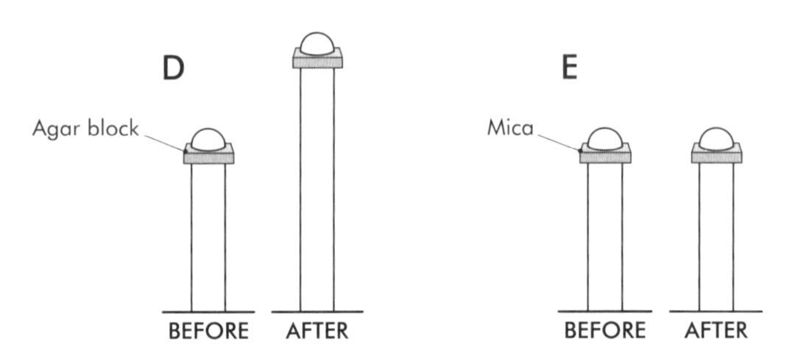

Fig. 24.3

Batch A	No treatment.
Batch B	Coleoptile tip cut off and immediately replaced on the cut surface of the coleoptile.
Batch C	Coleoptile tip cut off and discarded.
Batch D	Coleoptile tip cut off, a thin block of agar was placed on the cut surface of the coleoptile and the coleoptile tip then placed on top of the agar.
Batch E	Coleoptile tip cut off, a thin piece of mica (which is impermeable) was placed on top of the cut coleoptile and the coleoptile tip was then placed on top of the mica.

All of the batches of coleoptiles were grown in light from above for a period of time. Use your knowledge of biology to explain the results as shown in each of the above diagrams A–E.

chapter 25 *Animal Response to Stimuli*

SECTION A

1. (a) What is the CNS? ______________________

(b) What structures make up the CNS? ______________________

(c) What is the PNS? ______________________

(d) What are the components of the PNS? ______________________

(e) What is the name given to a nerve cell? ______________________

2. Label the following diagram of a neurone (Fig. 25.1).

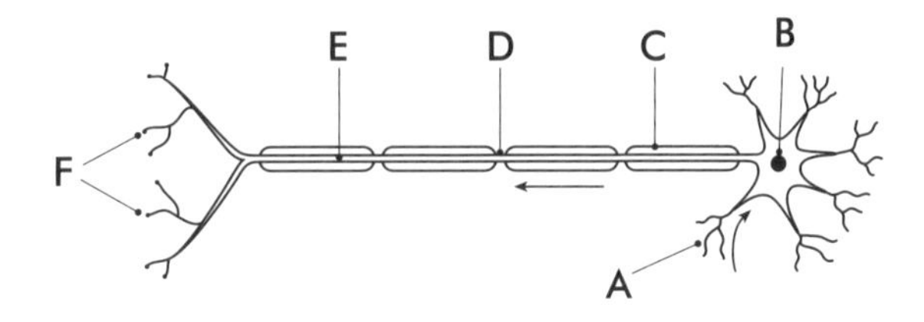

Fig. 25.1

A ______________ B ______________ C ______________

D ______________ E ______________ F ______________

3. Give the functions of the following parts of a neurone.

(a) Cell body ______________________

(b) Axon ______________________

(c) Dendrite ______________________

(d) Myelin sheath ______________________

(e) Schwann cell ______________________

(f) Synaptic knob ______________________

4. (a) Name the three main types of neurone found in humans and give the function and the difference in structure of each type.

(i) ______________________

____

____

(ii) ______________________

____

____

(iii) ______

(b) A nervous message is sometimes described as an 'all or nothing' message. Why do you think this is a good way of describing the type of message sent by neurones?

______

(c) Describe how chemicals have a role in the transfer of nervous messages.

______

(d) Why is there a delay between one message and the next?

______

(e) What happens to a message when it comes to the end of a neurone?

______

5. (a) What is a neurotransmitter? ______

(b) How does cocaine affect the nervous system?

______

(c) Why do cocaine addicts find that they need to take more and more of the drug to get the initial effect?

______

6. (a) What causes Parkinson's disease?

______

(b) What are the main symptoms of this disease?

______

(c) How can the symptoms be relieved?

____

____

(d) Why does physical damage to a spinal cord often lead to paralysis?

____

____

7. (a) What are the meninges?

____

(b) What is the white matter made up of in the CNS?

____

(c) What is the difference between the white matter and the grey matter?

____

____

(d) Where is cerebrospinal fluid found?

____

____

(e) What develops in the embryo from swellings at one end of the spinal cord?______

____

8. Name the labelled parts in the diagram of a vertical section through a human brain (Fig. 25.2).

A G B F C E D

Fig. 25.2

A ______________________________

B ______________________________

C ______________________________

D ______________________________

E ______________________________

F ______________________________

G ______________________________

Give the function(s) of the parts labelled:

A ______________________________

B ______________________________

C ______________________________

D ______________________________

E ______________________________

F ______________________________

G ______________________________

9. (a) Name the parts labelled in the following diagram of a cross section of a spinal cord (Fig. 25.3).

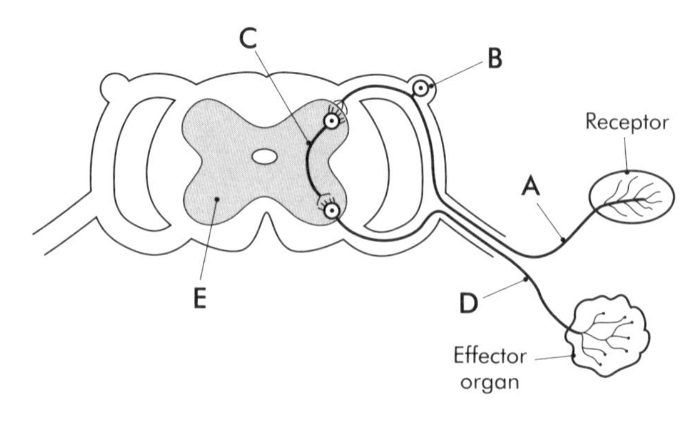

Fig. 25.3

A ______________________ B ______________________

C ______________________ D ______________________

E ______________________

(b) What protects the outside of the spinal cord? ______________________

(c) What is present in the hollow centre of the spinal cord? ______________________

(d) What is a reflex action? ______________________

____

(e) Describe how a simple reflex action works.

____

____

____

(f) What is the benefit of having reflex actions?

____

SECTION C

1. (a) Using labelled diagrams compare the structure and function of motor and sensory neurones.
(b) Most animals have a nervous system but plants do not. Why?
(c) The nervous system is divided into two main components. Name these and describe the functions of each part.

2. (a) Describe how messages are sent along a neurone and between neurones.
(b) Cocaine blocks the normal breakdown of a neurotransmitter. Explain how this gives cocaine its effect on humans.
(c) The drug curare is a muscle relaxant, which stops the normal contraction of voluntary muscles. The drug has a shape very similar to acetylcholine and is found on the receptors for acetylcholine in muscles, when it is administered to humans. Can you suggest how it has its effect?

3. (a) Describe a disorder of the nervous system due to a problem with a neurotransmitter and describe how its effects can be reduced.
(b) Describe the structure and function of the CNS.
(c) Why does physical damage to a spinal cord have such a widespread effect on a person?

4. (a) List the main areas of the brain and give their functions.
(b) There is a saying: 'Left-handed people are the only right-minded people around'. Comment on the validity of this statement.
(c) Describe how the brain has a major influence on the hormonal system.

5. John lifted his dinner plate out of the oven. He immediately let go of the hot plate yelling, but failed in his attempt to stop it hitting the floor, and so his dinner fed the dog.
(a) What is a reflex action?
(b) How does John's initial response demonstrate such an action.
(c) Describe how receptors, neurones and muscles work together to give a reflex action.
(d) What happened to cause John's attempt at rescuing his dinner?
(e) What might be the benefits of such reactions to John?

chapter 26 *Reception of Stimuli*

SECTION A

1. Fill in the blanks in the following table.

Name of Receptor	Function	Location
Interoreceptors		Arteries
	Respond to changes in external environment	
Chemoreceptors		Nose
	Respond to physical change	
		Eye
Thermoreceptors		

2. (a) Why are the senses of taste and smell considered to be similar?

____

(b) The nose does most tasting. Why is this?

____

____

(c) Where is the sense of touch localised? ____________________

(d) The receptors to what stimuli are considered to be part of the sense of touch?

____

____

(e) Why do you think the loss of the sense of pain is considered very dangerous?

____

____

3. (a) Name the parts labelled in the diagram of the ear (Fig. 26.1).

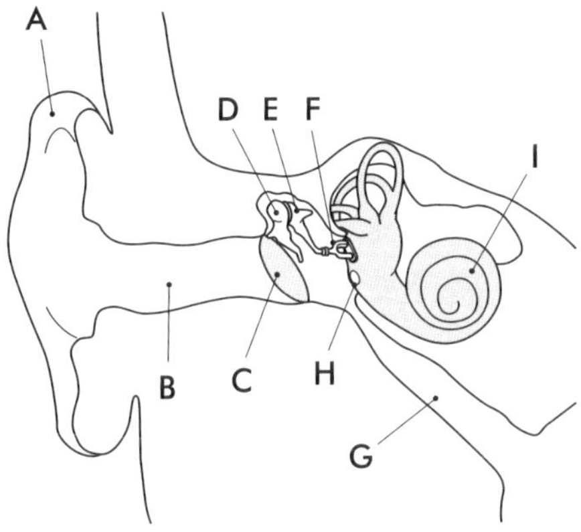

Fig. 26.1

A ______________________________

B ______________________________

C ______________________________

D ______________________________

E ______________________________

F ______________________________

G ______________________________

H ______________________________

I ______________________________

(b) (i) What is the function of C?

____

____

(ii) Why is this necessary?

____

____

(c) What is the function of the ear ossicles?

____

(d) (i) What structures are found in the cochlea?

____

(ii) What is the function of these structures?

____

(iii) What is the result of damage to these structures?

____

4. Name the labelled structures in the diagram of an eye and give a function for each structure (Fig. 26.2).

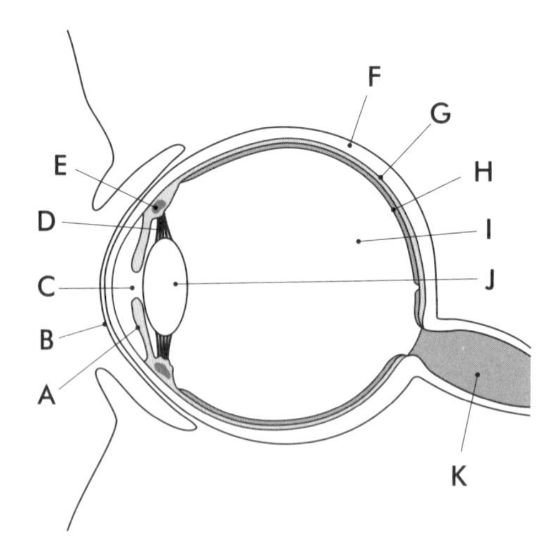

Fig. 26.2

A ______________ ________________________________

B ______________ ________________________________

C ______________ ________________________________

D ______________ ________________________________

E ______________ ________________________________

F ______________ ________________________________

G ______________ ________________________________

H ______________ ________________________________

I ______________ ________________________________

J ______________ ________________________________

K ______________ ________________________________

5. (a) What is the function of the fovea (yellow spot)?

____

(b) What is found at the fovea to carry out this function?

____

(c) (i) What is missing at the blind spot?

____

(ii) What effect does this have on the eye?

____

(iii) Why is this not normally noticed?

____

(d) What happens to light when it hits the lens and cornea?

____

6. (a) In the following diagrams name the eye defect shown and the type of lens you would use to correct the defect (Fig. 26.3).

Fig. 26.3

A Defect ________________ B Defect ________________

Lens ________________ Lens ________________

(b) (i) What is wrong with the eye to cause defect A?

____

(ii) What does this cause to happen when focusing?

____

(c) (i) What is wrong with the eye to cause defect B?

____

(ii) What does this cause to happen when focusing?

____

SECTION C

1. Give a biological explanation for the following:

(a) Passengers on planes are recommended to suck sweets on take-off or landing.

(b) On getting a cold you can lose your taste for food.

(c) Animals who are hunters have their two eyes at the front of their face, but their prey have them on the side of their face.

(d) If you get an earache you can feel dizzy.

2. The graph below shows the numbers of rods and cones on the retina of a human (Fig. 26.4).

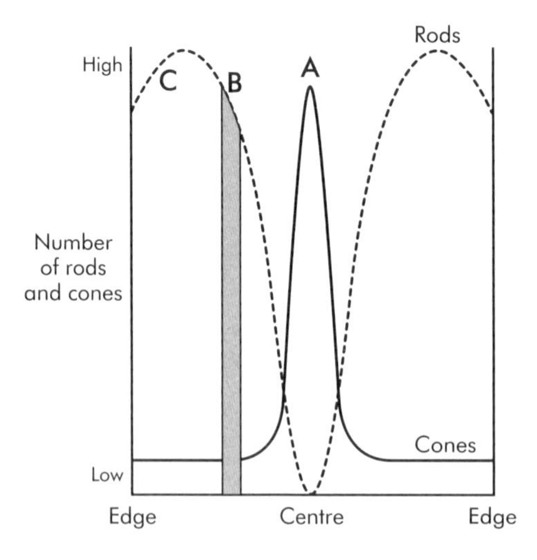

Fig. 26.4

(a) Name the part of the retina labelled A and B. Explain why there is a part of the eye that has no receptors.

(b) Why is part A the area of most accurate vision in the eye?

(c) Why is part C the area of the eye used in dim light?

(d) Which part of the eye is used for colour vision and why?

(e) Which part of the eye is likely to be most affected by a lack of vitamin A and why would this be the case?

3. (a) Explain what causes deafness to occur.

(b) What is the second function of the ear and what structures are present in the ear to carry out this function?

chapter 27 *The Endocrine System*

SECTION A

1. The endocrine system produces ______________, which are ______________ messages produced in one part of the body that causes a ______________ in another part of the body. Hormones are produced in ______________ glands which have no ______________ leaving them to carry the hormones, rather they ______________ their hormones directly into the ______________.

2. Complete the following table.

Hormone	Site of production	Site/s of action	Function
Growth			
	Thyroid gland		
			Control blood sugar level
Adrenaline			
	Testis		
Oestrogen			

3. (a) Where is the growth hormone produced?______________

(b) What is the function of this hormone in the body?

____

(c) What would be the effect of too low a level of this hormone?

____

(d) How could this problem be reduced?

____

(e) When must this treatment be given and why?

____

____

(f) What might cause an overproduction of this hormone?

____

4. In the following diagram of the human body indicate the location of the main endocrine glands (Fig. 27.1).

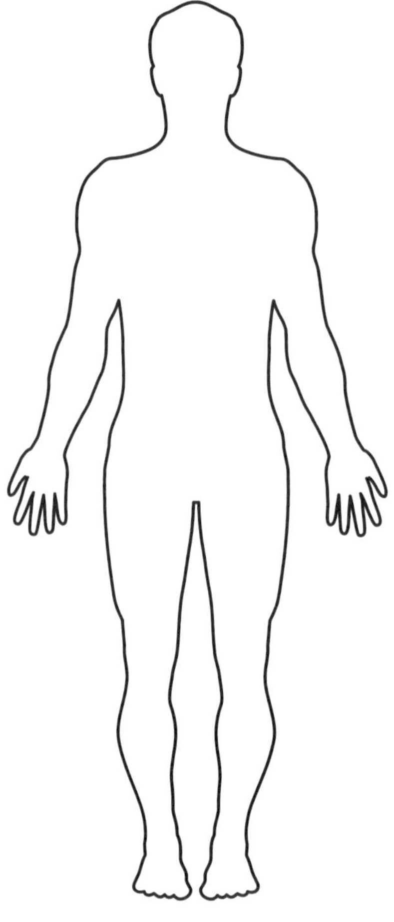

Fig. 27.1

5. (a) In the space below draw a diagram to illustrate the control process used to regulate the production of thyroxine.

(b) What produces the hormone TRH? ______________________________

(c) When is TRH produced?

____

(d) What gland does TRH affect? ____________________

(e) What is the function of TRH?

____

(f) What gland does TSH affect?

____

(h) What is the effect of TSH?

____

(h) What monitors the blood for the presence of thyroxine? ____________________

(i) What effect does thyroxine have on this tissue?

____

SECTION C

1. (a) Draw an outline diagram of the human body to show the location of the main endocrine glands.

(b) In the case of four of these glands name the hormone produced and give its function/s.

(c) Indicate the problems associated with the over- and underproduction of one of these hormones and suggest how the problems might be overcome.

2. (a) Compare the hormonal system with that of the nervous system.

(b) Why do animals need both types of systems?

(c) What are the two types of hormones and how do they affect their target cells?

(d) What is the role of the hypothalamus and the pituitary in the endocrine system?

3. High levels of blood sugar are dangerous to a person. There is a mechanism by which excess blood sugar can be taken out of the blood and temporarily stored in the liver as glycogen. In an experiment a person was given a glucose drink. For the next hour the level of glucose and insulin in the blood of the person was monitored. The results are shown in the graph below (Fig. 27.2).

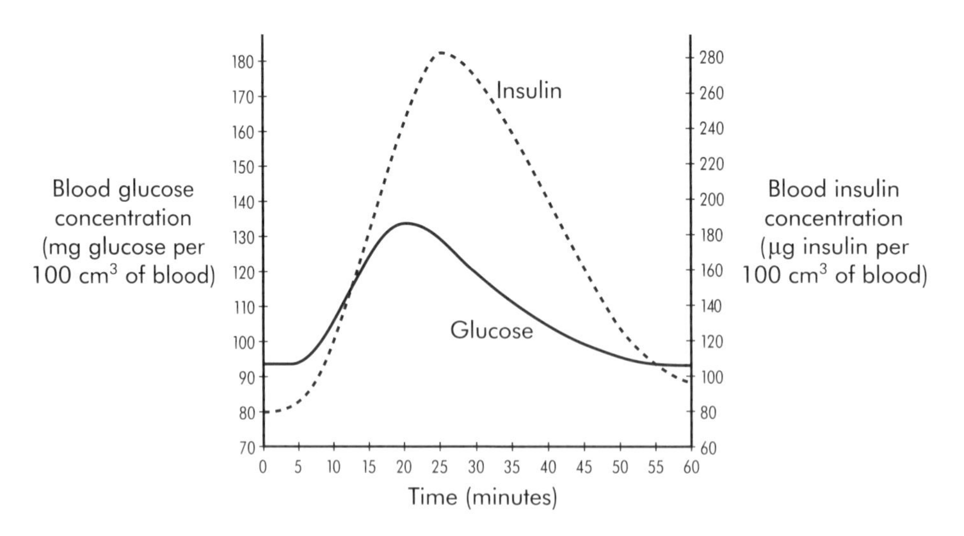

Fig. 27.2

(a) Why is there a delay between the person drinking and the rise in sugar level?
(b) Using the information given above and in the graph describe the function of insulin.
(c) Why does the level of insulin fall after 30 minutes?
(d) Using the graph suggest the normal level of glucose in the blood of a human.

4. (a) Frog tadpoles go through a process of metamorphoses (change in structure) to turn into adult frogs. A series of experiments was carried out using tadpoles.

(i) Tadpoles in iodine-free water did not change into frogs.

(ii) Tadpoles that were given thyroxine or extracts from a thyroid gland changed into frogs.

(iii) Tadpoles, which had the pituitary gland or their thyroid gland removed, failed to develop into frogs.

(iv) Tadpoles, which had their thyroid gland removed but were given thyroxine, changed into frogs.

From your knowledge of the hormone system explain what controls the metamorphosing of tadpoles into frogs. Back up your theory with evidence from the experiment.

5. (a) Describe how the level of the hormone thyroxine is controlled in humans.
(b) Why is this control process necessary?

chapter 28 *The Musculoskeletal System*

SECTION A

1. (a) Give the four functions of the skeleton.

___

___

___

___

(b) What is the function of the following structures?

(i) Cartilage. ___

(ii) Tendons. ___

(iii) Ligaments. ___

(iv) Muscles. ___

2. (a) The human skeleton is divided into two parts known as:

A. ___ made up of ___

B. ___ made up of ___

(b) Name the regions of the vertebral column.

___

(c) What type of joint is found between each vertebra?

___

(d) (i) What structure is found between each vertebra?

___

(ii) What is the function of this structure?

___

(e) What are the functions of the ribs?

___

3. (a) In the space provided draw a simple diagram to show the arrangement of bones in a pentadactyl limb.

(b) This structure can be used as evidence for evolution. Why is this?

____

____

____

4. (a) Label the following diagram of a leg bone as seen in longitudinal section (Fig. 28.1).

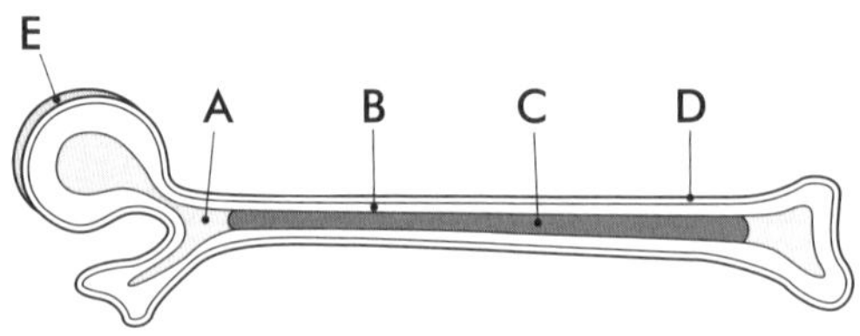

E A B C D

Fig. 28.1

A ____________ B ____________ C ____________

D ____________ E ____________

(b) What is found in part A and what is its function?

____

____

(c) What is the function of part E?

____

(d) What is the function of part D?

____

(e) Why is the bone not solid?

____

(f) What is the role of the sun in bone manufacture?

____

5. (a) In the space provided draw a simple diagram of the structure of a synovial joint and give a function for each labelled part.

(b) Name the four types of synovial joints and give an example of each type.

______________________________ ______________________________

______________________________ ______________________________

______________________________ ______________________________

______________________________ ______________________________

(c) There are two other types of joints. Name them and give an example of each type.

______________________________ ______________________________

______________________________ ______________________________

6. (a) Motor nerve, ligaments, tendons, biceps and triceps. All of these structures are used when moving the arm. Describe how these structures work together to carry out this function.

____

____

____

____

____

(b) What are antagonistic muscles?

(c) Why can the triceps and the biceps be considered antagonistic muscles?

(d) Why is it necessary to have voluntary muscles in antagonistic pairs?

7. (a) What is meant by the term autoimmune disease?

(b) Why can rheumatoid arthritis be considered such a disease?

(c) What is the difference between rheumatoid arthritis and osteoarthritis?

(d) What is osteoporosis?

(e) What steps can be taken to reduce the damage of osteoporosis?

8. (a) What is ossification?

(b) Which cells produce ossification?

(c) What happens to stop ossification?

(d) Why must growth hormone, if medically required, be given before ossification stops?

9. (a) What is the difference between the following pairs of cells?

(i) Osteoblasts and osteocytes.

____

____

(ii) Osteoblasts and osteoclasts.

____

____

(b) What is the problem of weight-free growth of bones?

____

____

(c) What are the differences between the structure of bone and cartilage?

____

____

SECTION B

1. (a) Draw a well-labelled diagram to show a longitudinal section through a typical long bone. Give a function for each part labelled.

(b) Describe the different types of joints found in the human skeleton giving named examples.

(c) Give an illustrated account of the process used to move the arm bones.

2. (a) With the aid of a diagram describe the structure and function of a synovial joint.

(b) There are other less flexible joints in the skeleton. Where are these found and what is their function?

(c) Describe two skeletal disorders and suggest ways in which their symptoms can be reduced.

3. (a) Describe the process of ossification.

(b) After ossification the bones of an adult begin to loose calcium and this reduces the bone density of an adult over time. It has been shown that the presence of the sex hormones has a major effect on this process.

If we take a 20-year-old person as having 100 per cent bone density then we can chart the loss of bone density in the remainder of a person's life. This is shown in the graph below (Fig. 28.2).

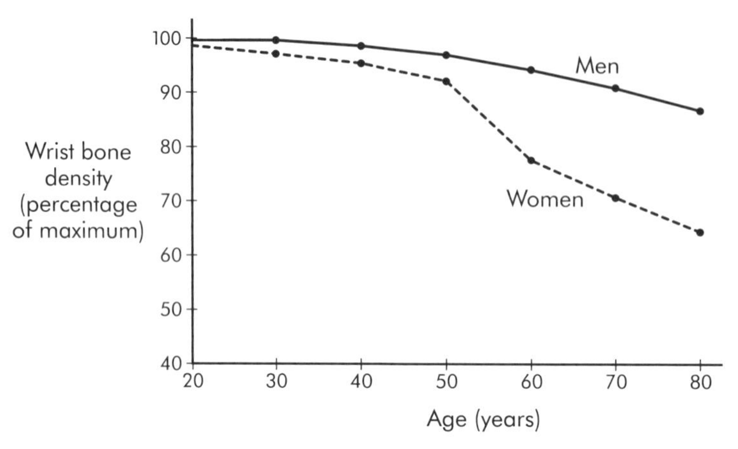

Fig. 28.2

(i) What happens to a woman after the age of 50 to make her lose more bone density than a man?

(ii) Can you suggest any way in which loss of bone density can be reduced?

(iii) There is some evidence to suggest that certain older people (particularly women) break their femur and then fall rather than the other way around. Is there any evidence in the graph to suggest a reason for this?

4. (a) In an experiment a chicken bone was put into HCl (this acid will remove inorganic salts from the bone). At the end of the experiment the bone was discovered to be very flexible and not at all hard. In a second experiment a chicken bone was burnt (burning will remove all organic material). At the end of this experiment the bone was found to be very brittle and not at all flexible. These experiments were used to suggest that the matrix of bone is made up of two materials. Do you agree with this? Give reasons for your answer.

(b) Describe the process by which bones are constantly being made and broken down in humans.

(c) Astronauts and bedridden patients tend to loose bone density. Why is this?

chapter 29 *The Human Defence System*

SECTION A

1. (a) What is a pathogen? ______

(b) What is the function of the general defence system?

______

(c) In what way does the specific defence system differ from this?

______

(d) What is the advantage of the specific defence system?

______

______

2. (a) Describe four ways in which bacteria are prevented from gaining access to the body.

______

______

______

______

(b) What happens in the inflammatory system?

______

______

______

(c) What is the function of phagocytic white blood cells?

______

(d) What is the role of complement in the defence system?

______

(e) What must happen to cells before they produce interferon?

______

(f) What is the function of interferon?

______

3. (a) What is the other name for the immune response? ______

(b) What is the difference between an antigen and an antibody?

______

______

(c) What produces antibodies? ______

(d) Why is this process called the specific defence system?

______

(e) What must happen before the body can produce specific antibodies against an antigen?

______

______

(f) What happens after an antibody has destroyed the micro-organism?

______

______

4. (a) What is induced immunity?

______

______

(b) What is the difference between natural and artificial immunity?

______

______

(c) What happens in passive immunity?

______

______

(d) What might be an advantage of this type of immunity?

______

______

(e) Describe how a person can get active immunity.

______

______

(f) Why is it that a human normally only gets an infectious disease once?

______

______

5. (a) What is a virus? ______

(b) Why are viruses not considered to be living organisms?

______

(c) While studying the tobacco mosaic disease in 1892, the experimenters filtered the plant extract through a filter fine enough to remove any bacteria. They were very surprised to find the filtrate still caused the disease. Why do you think this was the case?

____

____

____

(d) What instrument was needed before scientists could see the virus?

____

(e) What part of the virus allows it to attach to the target cells?

____

6. (a) What cell organelle is the target of a virus? ______________________

(b) What processes occur when the virus causes cell lysis?

____

____

____

(c) What is the other effect the virus can have on the cell?

____

____

(d) What could be the long-term result of this second effect?

____

____

7. (a) What is the function of the B lympocytes?

____

____

(b) What is a clone?

____

(c) What is the difference between each B lympocyte clone?

____

(d) (i) What is attached to the surface of the B lympocyte? ______________

(ii) What is the function of this structure?

____

____

(e) (i) What are memory B cells?

(ii) Why are they so important in active immunity?

8. (a) Name the three different types of T lympocytes and give a function of each of these.

(i)

(ii)

(iii)

(b) What is the function of the T memory cells?

(c) How does HIV effect the T cells?

(d) Explain how HIV causes AIDS.

SECTION C

1. (a) Describe how the general defence system of the human carries out its function.

(b) Explain how the entry of a foreign substance into the body stimulates the specific immune system into action.

(c) Why is this second defence system necessary?

2. (a) Acquired immunity can be active or passive and artificial or natural. Describe how each of these is acquired and the benefit of each of them.

(b) The specific defence system is often considered to be the more important of the two defence systems in the human. Do you agree? Give your reasons.

(c) When a patient is getting chemotherapy or when a person is suffering from AIDS their defence system is damaged. These patients are open to a lot of infections due to micro-organisms already present in the body. Why do you think this is the case?

3. (a) Give an illustrated account of the structure of viruses.
(b) Describe the effect of viruses on cells.
(c) What role is played by viruses in genetic engineering?

4. In the final years of the 19th century, Dublin was hit by a major outbreak of typhoid fever. A resident of Dublin, Mary Mallon, lost her entire family and emigrated to the USA. Mary became an excellent cook and was much in demand. However, she was dogged by bad luck and after working for a few months in any one household, typhoid would break out. Mary would move on to other employment (with very good references). In 1906 when typhoid broke out in General Warren's house, the cases were investigated as it was unusual for typhoid to be found in wealthy people's households since good sanitation was available.

The only common link in all the cases was Mary Mallon and she was discovered to be a 'healthy carrier'. This was the first time this condition was identified. Mary was originally kept at a hospital but later released on condition that she did not return to cooking. This, however, she did and another outbreak was discovered in a hospital where she was working. This time the public got to hear of the effect of Mary, who was called 'Typhoid Mary' in the press. She was nearly lynched by an angry mob. As it was impossible to get rid of the typhoid microbes from Mary, she spent the rest of her life living in a house attached to the hospital.

(a) What is a healthy carrier?
(b) How do you think Mary passed on the disease?
(c) Describe the mechanism by which most people recover from a case of typhoid.
(d) What measures could Mary have taken to reduce the likelihood of passing on the disease?
(e) How would doctors solve the problem of a typhoid carrier today?
(f) If you were trying to trace the source of the outbreak in 1906 in a scientific manner, what would you have done?

5. (a) Describe how B lympocytes function and how they confer active immunity.
(b) Explain how T lympocytes function in the control of the immune system?

6. The Human Immunodeficiency Virus (HIV) eventually causes AIDS. In this disease the human immune system fails to work and the patient is eventually overcome by a microbial infection. This retrovirus has the ability, along with influenza and the common cold, to continuously change its genetic code for the proteins that are expressed on the outside of the virus. In the case of HIV the virus lives in the T helper cells and eventually destroys them.

(i) Explain how the ability to change its proteins makes it impossible for our specific defence system to protect us from recurring bouts of flu or the common cold.

(ii) In the case of HIV explain why the immune system cannot destroy the virus, and secondly, how it eventually leads to AIDS.

(iii) How can you reduce the risk of getting HIV?

chapter 30 *Vegetative Reproduction in Plants*

SECTION A

1. 1. Answer the following by placing a (✓) in the correct box in each case.

(a) Vegetative reproduction:

Involves fusion of gametes	☐	Involves the production of seeds	☐
Is a method of asexual reproduction	☐	Is a method of sexual reproduction	☐

(b) Clones are:

Produced by sexual reproduction	☐	A breed of sheep	☐
Able to store food	☐	All genetically identical	☐

(c) A bulb is:

A modified stem	☐	A modified bud	☐
A modified flower	☐	A modified root	☐

(d) An example of a root tuber is found in the:

Potato	☐	Carrot	☐
Onion	☐	Dahlia	☐

(e) The following are all methods of artificial vegetative propagation except:

Grafting	☐	Producing runners	☐
Micropropagation	☐	Taking cuttings	☐

2. The diagram shows a vertical section through a bulb (Fig. 30.1).

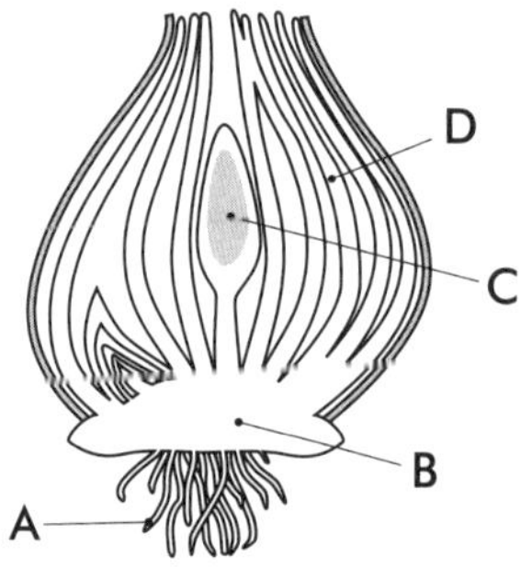

Fig. 30.1

(a) Name the parts labelled:

A ______________________ B ______________________

C ______________________ D ______________________

(b) Which of the following plants form a bulb?

Buttercup Dahlia Tulip Potato ______________________

(c) A bulb is used by some plants as a means of vegetative propagation. Briefly describe how a bulb carries out this function.

____

____

(d) Other than as a means of reproduction what other function does a bulb serve for the plant?

____

3. (a) List two advantages of vegetative propagation.

____

____

(b) List two disadvantages of vegetative propagation.

____

____

(c) List two advantages of reproduction by seed.

____

____

(d) List two disadvantages of reproduction by seed.

____

____

SECTION C

1. (a) Give the meaning of the term vegetative propagation.

(b) Describe one method of vegetative propagation.

(c) Suggest a reason why plant growers sometimes choose to propagate a particular plant by means of vegetative propagation rather than by seed.

2. Give an illustrated account of how vegetative propagation takes place in a named flowering plant. State two advantages and two disadvantages of this type of reproduction in plants.

3. Read the passage below about the micropropagation of plants, and answer the questions that follow.

<u>Micropropagation</u> is a technique that uses plant tissue culture to generate whole plants. Plant tissue culture involves growing single cells or tissues in controlled, <u>aseptic conditions</u>. Cells from the <u>meristems</u> of plants are grown in media

containing nutrients and growth regulators. The cells divide by mitosis to form new little plants. The tissue is first sterilised by rinsing in a dilute hypochlorite solution, it is next rinsed in sterile water and then placed on the sterile culture medium. The explants, as these meristem pieces are called, are kept in sealed petri dishes in a room where the conditions of light and temperature are strictly controlled. Explants can be almost any part of a plant, stem, root or leaf. The technique of micropropagation is used commercially to produce pot plants and cut flowers on a large scale. It is also used to propagate rare and endangered species that are not easily propagated by the usual plant breeding methods.

(a) Explain the underlined terms as used in the passage.

(b) Explain why the new plants produced by tissue culture form a clone.

(c) State two advantages to commercial growers of this method of propagation.

(d) The diagram below shows some steps in the micropropagation of a crop plant (Fig. 30.2).

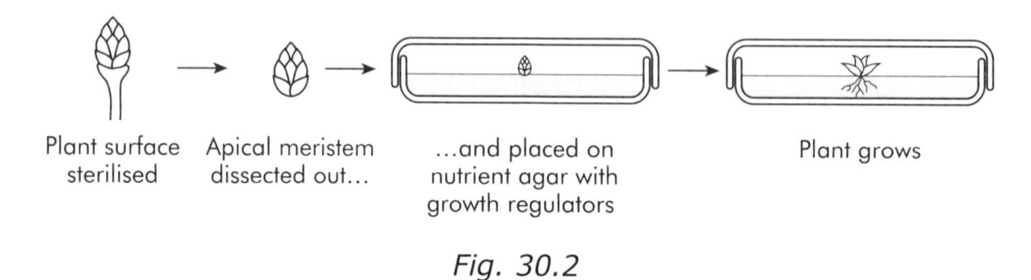

Fig. 30.2

(i) Name a plant tissue suitable for use in micropropagation.

(ii) What would you use to sterilise the surface of the tissue at step 1?

(iii) Give two precautions that should be taken to prevent contamination of the explants after their surface has been sterilised.

chapter 31 *Sexual Reproduction in Flowering Plants*

SECTION A

1. Answer the following by placing a tick (✓) in the appropriate box.

(a) The reproductive organ of the flowering plant is the:

stem ☐ root ☐ leaf ☐ flower ☐

(b) Part of the stamen:

stigma ☐ anther ☐ ovary ☐ style ☐

(c) Contains the ovules:

stigma ☐ anther ☐ ovary ☐ style ☐

(d) Protects the flower before it blossoms:

sepal ☐ petal ☐ anther ☐ carpel ☐

(e) The function of the anther is to:

produce nectar ☐ produce eggs ☐
produce scent ☐ produce pollen ☐

2. (a) Name the parts labelled P, Q, R, S, T on the diagram of the flower below (Fig. 31.1).

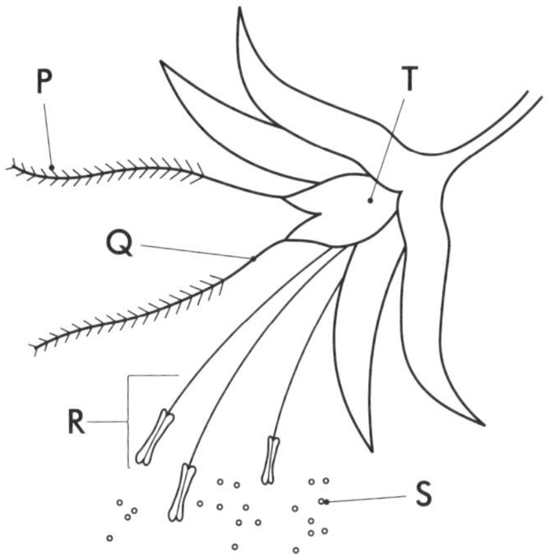

Fig. 31.1

P ______________________________

Q ______________________________

R ______________________________

S ______________________________

T ______________________________

(b) Suggest the method by which this flower is pollinated.

____

(c) State three features shown by the flower in Fig. 31.1 to support your answer to (b).

(i) ______________________________

(ii) ______________________________

(iii) ______________________________

3. Self pollination involves the transfer of pollen from the ____________ to the ____________ of the ______________ flower. Cross pollination involves the transfer of pollen from the _______________ to the ________________ of a ______________ flower.

Why is cross pollination considered better than self pollination?

____

4. The diagram shows the carpel of a flower before fertilisation takes place (Fig. 31.2).

(a) Name the parts labelled:

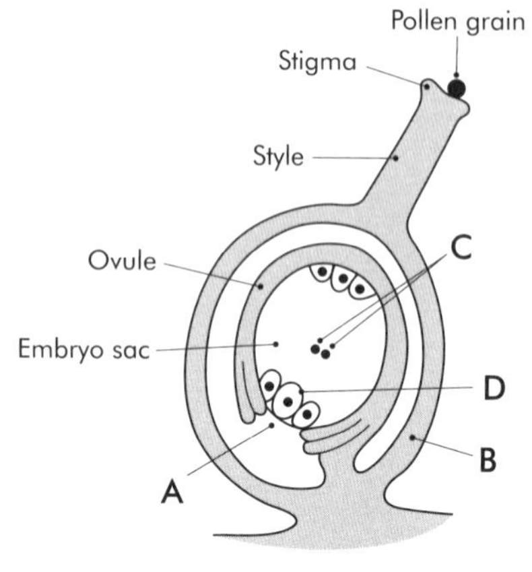

Fig 31.2

A ____

B ____

C ____

D ____

(b) What does the pollen grain produce?

____

(c) Is D haploid or diploid? ____

(d) Complete the diagram to show what happens to the pollen grain to allow fertilisation to occur.

5. (a) Label the following parts of the broad bean seed shown, using clear lines and the appropriate letter (Fig. 31.3).

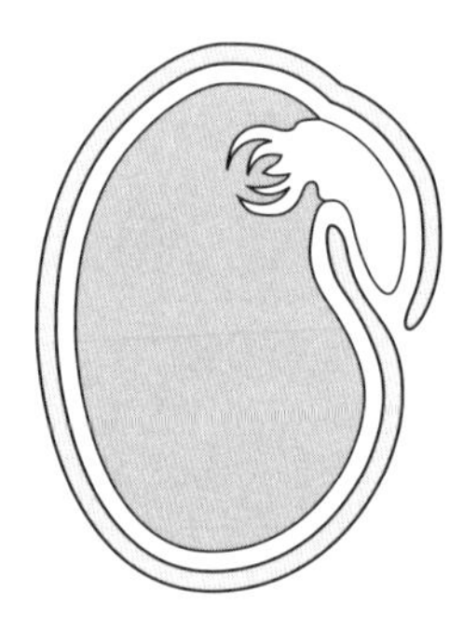

Fig. 31.3

Part	Letter
Plumule	A
Cotyledon	B
Radicle	C
Testa	D

(b) Match the word in the left-hand column with the correct description in the right-hand column.

Testa	stores food
Plumule	protects the seed
Embryo	the young root
Cotyledon	the plumule and radicle
Radicle	the young shoot

6. (a) State what the ovule of a flowering plant becomes following fertilisation.

____

(b) What is a cotyledon?

____

(c) Name a non-endospermic seed. ______________________________

(d) Why is water needed for germination?

____

____

(e) What name is given to the period of time when a seed remains viable but does not develop?

____

(f) Name a chemical that absorbs oxygen gas. ______________________________

(g) What do developing seeds produce to stimulate growth of fruit tissues?

____

7. Complete the following table.

Named fruit	**Method of dispersal**
Sycamore	
	Wind
Garden pea	
Blackberry	
	Self dispersal

8. Sexual reproduction in flowering plants involves different events including:

(a) Pollination which is the ____________ of ________________ from the _____________ of a stamen to the stigma of a _____________.

There are two main types of pollination. _______________ pollination which is helped by insects or by ______________ and ____________ pollination. Flowers that are insect-pollinated are specially adapted to attract insects.

Give the name of an insect-pollinated flower and state two features of the named flower that show adaptation to pollination by insects.

____

(b) Fertilisation follows pollination. Fertilisation is the ____________ of a pollen grain nucleus with an _____________ to form a __________________. Following fertilisation the ovule develops into a ___________ and the wall of the ovary becomes the _______________.

(c) Seeds and fruits are dispersed away from the parent plants to prevent ______________________. Seeds can be dispersed by __________________, ______________________ and ______________________.

(d) Following dispersal, seeds often pass through a period of ______________________ during which the immature embryo plant has time to _______________. If conditions are right the seed will germinate. The factors necessary for germination are ______________, ___________________ and ____________________. Finally a new seedling is produced which will grow and form new flowers and seeds.

9. The diagrams show the stages in the germination of a seed.

(a) Name and give the main function of each of the parts labelled A, B, C, D (Fig. 31.4).

Fig 31.4

Name		Function
A	________________	______________________________
B	________________	______________________________
C	________________	______________________________
D	________________	______________________________

(b) Which of the following condition(s) is not essential for the germination of most seeds? warmth light moisture oxygen

____

SECTION B

1. In the space provided below, draw a labelled diagram of the apparatus you used to show the digestive activity in seeds during germination.

Describe the steps you carried out in the activity, including your expected results and conclusions.

____

____

____

____

____

____

____

____

2. Give a scientific explanation for each of the following steps taken in the activity to show the effect of oxygen on the germination of seeds, such as cress seeds.

(a) Put some cotton wool into two test tubes and moisten it.

____

____

(b) Add 5–8 seeds to each test tube.

____

____

(c) Place a solution of alkaline pyrogallol into one of the test tubes.

____

____

(d) Place some water only into the other test tube.

(e) Leave the test tubes in the laboratory at room temperature.

(f) Leave the test tubes for one week.

3. You are asked to find the best temperature for germinating mustard seeds. You are given 80 seeds, 8 test tubes, some cotton wool, some water and a thermometer. In addition you have access to water baths, a refrigerator and an incubator (oven).

(a) In the space below draw and label one of the test tubes you would set up in the experiment.

(b) Describe the steps you would carry out in the experiment including setting up a control.

(c) Describe any one difficulty you might have in carrying out this activity.

____

____

____

SECTION C

1. (a) Draw a large labelled diagram of a named insect-pollinated flower.
(b) Define the term pollination.
(c) List three features of the flower you have drawn that show how it is adapted for insect pollination.

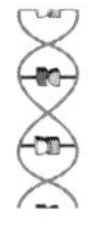

2. Describe with the aid of labelled diagrams the development of
(a) the pollen grain.
(b) the embryo sac in a named flowering plant.

3. (a) What is meant by dispersal of seeds?
(b) Give two reasons why seed dispersal is important for plants.
(c) Name the fruits shown in the diagrams below (Fig. 31.5).

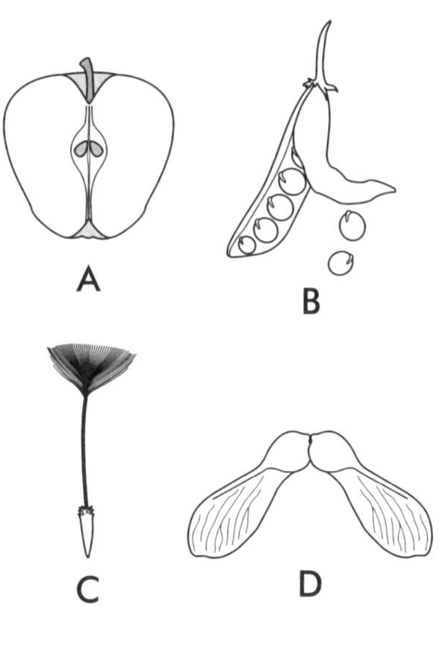

Fig. 31.5

(d) State which of the fruits shown is dispersed by (i) wind, and (ii) water.
(e) For each of fruits A and C describe two features that show adaptations to the method of dispersal.

4. The diagrams below show a section through a pea seed and a barley grain after they have both been stained with iodine solution (Fig. 31.6). The dark regions went blue/black when iodine solution was applied.

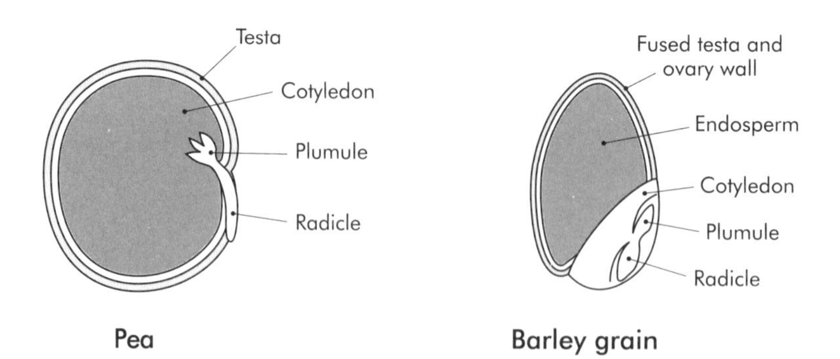

Fig. 31.6

(a) What is the main food material stored in the pea seed?
(b) What evidence from the diagram do you have to support your answer to (a)?
(c) What part of the pea seed stores this food material?
(d) What part of the barley grain stores this food material?
(e) What other food materials might be stored in the barley grain? Describe how you would test for these materials.

5. (a) Define the term germination.
(b) Water, oxygen and the correct temperature are needed for the successful germination of seeds. Describe the role of each of these factors in germination.

6. As seeds begin to germinate they produce enzymes such as amylase.
(a) Define the term enzyme and give an example other than amylase.
(b) Explain the role of enzymes in the germination of seeds.
(c) Describe an experiment you could carry out to show enzyme activity during germination of seeds.

7. The diagram shows a LS through a maize seed (Fig. 31.7).
(a) What does LS mean?
(b) Name the parts labelled R, S, T and U.
(c) Which parts represent the embryo plant?
(d) List the conditions necessary for seeds to germinate.
(e) What benefits are gained from keeping certain seeds in a 'seed bank'?

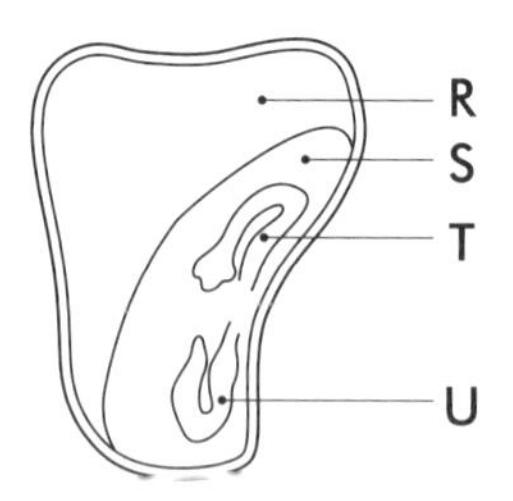

Fig. 31.7

8. (a) In what form is carbohydrate stored in (i) the human, and (ii) a broad bean seed? State where a carbohydrate store is to be found in (i) the human, and (ii) a broad bean seed.
(b) Maize is an example of an endospermic seed. Distinguish between an endospermic and a non-endospermic seed.

(c) In an experiment to study the changes in the lipid (fat) and carbohydrate content of the endosperm and embryo of germinating maize grains, the following results were obtained:

Time in days	**Content (grams per maize grain)**			
	Endosperm		**Embryo**	
	Lipid	**Carbohydrate**	**Lipid**	**Carbohydrate**
0	0.25	0.02	0	0.00
4	0.22	0.06	0	0.02
6	0.10	0.16	0	0.06
8	0.05	0.11	0	0.15
12	0.01	0.02	0	0.19

(i) Draw graphs using the same axes to show the changes in lipid and carbohydrate content with time. Put time on the horizontal axis.

(ii) Comment on the relationship, if any, between changes in the following and suggest an explanation: (1) the endosperm lipid and the endosperm carbohydrate; and (2) the endosperm carbohydrate and the embryo carbohydrate.

[HINT: read the graphs by the changes in the graph lines, and compare the changes.]

9. The graph below shows the growth curve of a plant from the time it germinates until it dies (Fig. 31.8). The fresh weight (weight of tissues with water) indicated at A is that of a seed from the plant some time after the seed has been dispersed. Examine the graph and answer the questions that follow:

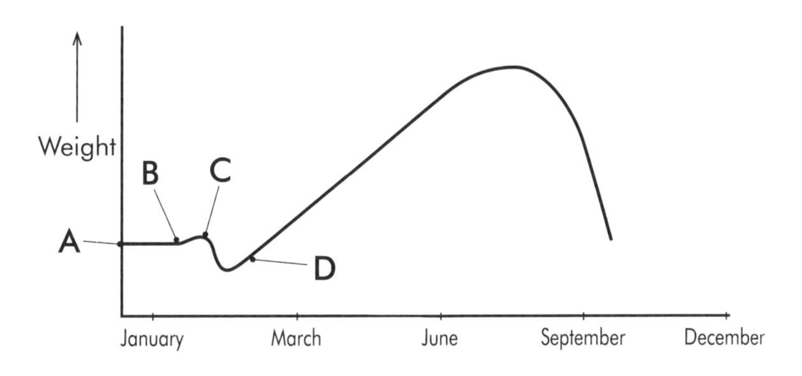

Fig. 31.8

(a) What term is used to describe the period of time in the plant's life cycle between stage A and B on the graph? State the importance of this period.

(b) Give the name for the stage in the life cycle of the plant that begins at B. Give one reason for (i) the increase in fresh weight between B and C, and (ii) the decrease in fresh weight between C and D.

(c) Fresh weight increases rapidly from D. Explain what is happening to account for this rise.

(d) Copy the graph and mark F on it where you think flowering of this plant is likely to begin, and G where seed dispersal is likely to occur.

chapter 32 Human Reproduction 1

SECTION A

1. (a) The organ in which sperm are made is the ______________________

(b) The male reproductive hormone is called ______________________

(c) Sperm cells are produced by a type of cell division called ______________

(d) Name the tube that transports sperm in the male. ______________

(d) What is the function of the prostate gland?

____

2. (a) The organ in which eggs are made is the ______________________

(b) The female reproductive hormones are ______________________

____

(c) Egg cells are produced by a type of cell division called ______________

(d) Name the tube down which the egg travels when it leaves the ovary.

____

(e) What is the function of the uterus?

____

3. Match the following parts of the male and female reproductive organs with the descriptions given.

Vagina	The urethra runs through the length of this organ
Epididymis	The neck of the womb
Seminal vesicle	It connects the testes to the urethra
Penis	The place where fertilisation takes place
Vas deferens	Large coiled tube which store sperm
Cervix	Produces fluid to allow sperm to move
Oviduct	Produces eggs
Ovary	Connects the uterus to the exterior

4. Complete the following passage using the terms listed below:

puberty hormones endometrium testes
shed female menstruation ovaries male

Sperm are produced in the ______________ of the ________________. The ________________ produce eggs which are the ______________ gamete. Chemicals called ____________________ control the process of reproduction in humans. Every month from ___________________ onwards a girl produces many potential eggs, and the lining of the womb, the ________________________ becomes built up in case fertilisation occurs. Usually, however, fertilisation does not occur and the lining of the womb is no longer needed and is ____________. This event is known as ________________________.

5. The diagram shows the human female reproductive system (Fig. 32.1).

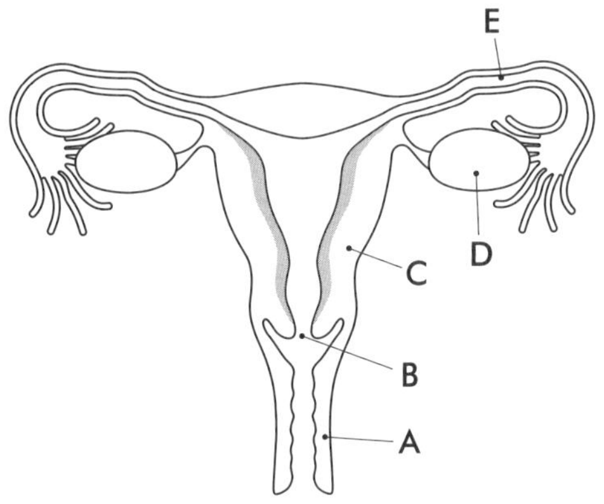

Fig. 32.1

(a) Name the parts A–E.

A ________________________________

B ________________________________

C ________________________________

D ________________________________

E ________________________________

(b) What is the endometrium? ________________________________

(c) What is the function of the endometrium? ________________________________

____

(d) Indicate on the diagram the location of the endometrium.

6. State whether the following statements are true or false. If a statement is false, give a reason for your answer.

(a) Sperm cells are a type of gamete. ______________________

____

(b) The time in a woman's life when she starts to produce eggs is known as the menopause.

____

____

(c) Ovulation is the release of an egg from an ovary. ______________________

____

(d) Testosterone is the male sex hormone. ______________________

____

(e) Progesterone is responsible for the secondary sexual characteristics in the female.

____

____

(f) Menstruation is 28 days long. ______________________

____

(g) The urethra connects the testis to the penis. ______________________

____

SECTION C

1. (a) Draw a large labelled diagram of the male reproductive system of the human.

(b) Name which part(s) of the system is/are responsible for each of the following: (i) production of semen, (ii) storage of sperm and (iii) release of sperm.

(c) Name the type of cell division that is used to produce sperm. Explain why this type of cell division is used to produce gametes.

(d) Suggest a reason why the testes are found in the scrotal sacs and not inside the body cavity.

(e) The testes, in addition to producing sperm, produce a male hormone. Name this hormone and describe the changes this hormone makes in a boy's body during puberty.

2. (a) Distinguish between the menstrual cycle and menstruation.

(b) What is the function of the menstrual cycle?

(c) Where does the menstrual cycle take place?

(d) Examine the diagram of a typical 28-day menstrual cycle shown below and answer the following (Fig. 32.2):

(i) Name the events marked A, B and C.

(ii) Between which days of the cycle shown do the following occur?

ovulation; menstruation; formation of follicle; breakdown of corpus luteum.

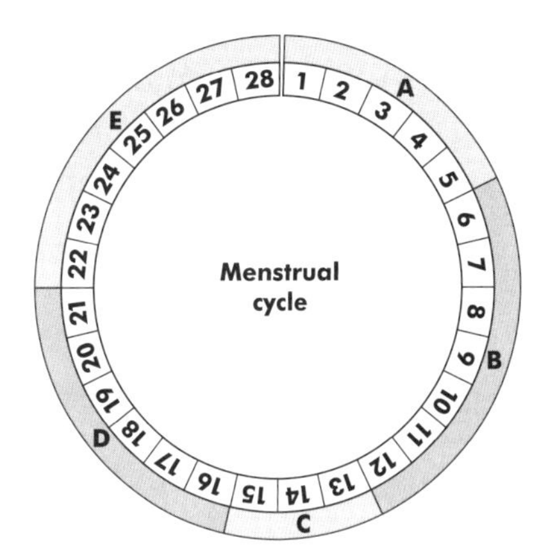

Fig. 32.2

(e) Name two hormones that control the menstrual cycle.

(f) Do all women have regular 28-day cycles like that described above? Explain.

3. The diagram below shows the relationship between the thickness of the uterus lining, ovulation and the levels of oestrogen and progesterone during two menstrual cycles (Fig. 32.3).

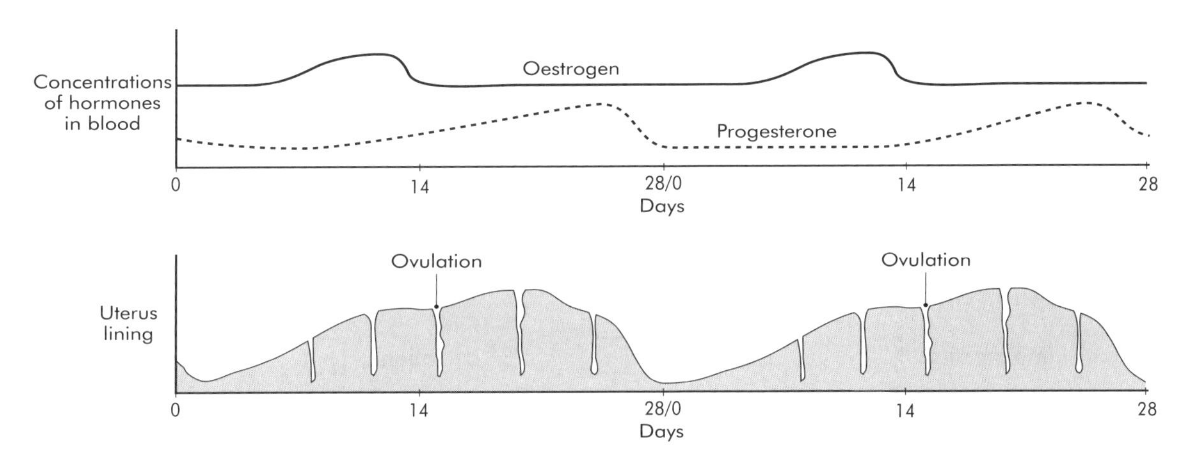

Fig. 32.3

(a) Explain the term ovulation.

(b) What is another name for the uterus lining?

(c) State the process that occurs between day 0 to day 5.

(d) State the changes that take place in the lining of the uterus from day 5 to day 28 of each cycle.

(e) What is the reason for these changes?

(f) Where and when are oestrogen and progesterone produced?

(g) What effect does the level of oestrogen have on (i) the thickening of the uterus lining, and (ii) ovulation?

(h) What effect does progesterone have on (i) the thickening of the uterus lining, and (ii) ovulation?

(i) Name two other hormones produced during the menstrual cycle and say where each of them is produced.

4. (a) What are fibroids?
(b) Where do they form?
(c) Why do they form?
(d) What is a suitable treatment for fibroids?

5. (a) What is endometriosis?
(b) How does endometriosis occur?
(c) Why does endometriosis occur?
(d) What is a suitable treatment for endometriosis?

chapter 33 *Human Reproduction 2*

SECTION A

1. Answer the following by placing a tick (✓) in the appropriate box.

(a) In humans fertilisation normally occurs in the:

uterus ☐ oviduct ☐ vagina ☐ ovary ☐

(b) A fertilised egg divides to form a ball of cells called a/an:

embryo ☐ amnion ☐ ovum ☐ foetus ☐

(c) Which of the following is the name given to the fertilised egg sinking into the lining of the uterus?

implantation ☐ gestation ☐ ovulation ☐ menstruation ☐

(d) The foetus is surrounded by a thin membrane called the:

placenta ☐ uterus ☐ amnion ☐ oviduct ☐

(e) In humans pregnancy lasts approximately:

28 days ☐ 40 weeks ☐ 28 weeks ☐ 40 days ☐

2. The diagram below shows the human female reproductive system (Fig. 33.1). Complete the table below by selecting a letter from the diagram to identify where each of the following events take place.

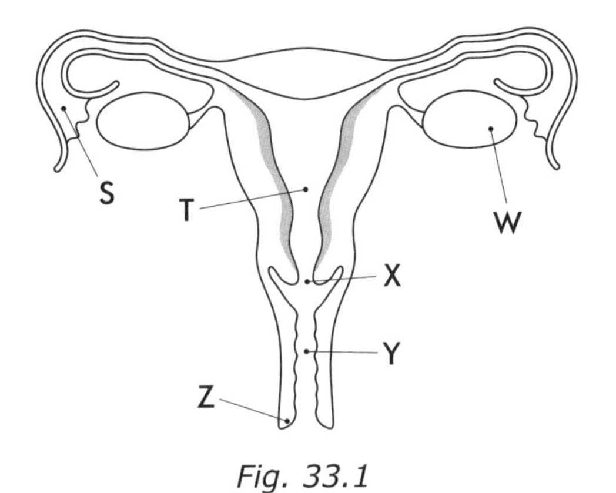

Fig. 33.1

Event	Letter
Ovulation	
Fertilisation	
Implantation	

3. The diagrams show the events from <u>fertilisation</u> at A to <u>implantation</u> at D (Fig. 33.2).

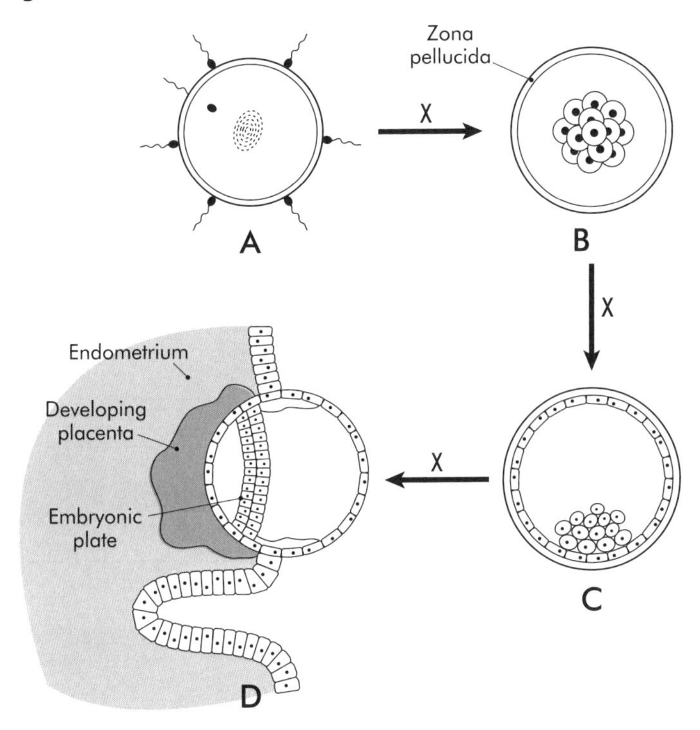

Fig 33.2

(a) Define the underlined terms as they apply to the human.

Fertilisation

____

____

Implantation

____

____

(b) Name the structure at B. ______________________

(c) Name the structure at C. ______________________

(d) Name the type of cell division at X. ______________________

(e) Name the part of the uterus into which C implants itself. ______________________

(f) Roughly how many days after fertilisation is implantation complete? ____________

4. State whether the following statements are true or false. If a statement is false, give a reason for your answer.

(a) In humans fertilisation normally occurs in the oviduct. ______________

____

(b) The release of sperm from the penis is called an erection. ____________

____

(c) A human egg can live for 28 days. ________________

____

(d) The fertile period is the time in the menstrual cycle when fertilisation cannot occur.

____

____

(e) Human pregnancy lasts for 40 weeks. ________________

____

(f) Testosterone is the male sex hormone. ________________

____

(g) The term 'test-tube baby' means the baby grows in a test tube. ____________

____

(h) A low sperm count is not a cause of male infertility. ________________

____

5. Read the sentences below and then re-write them in the order that best describes the events from ejaculation to implantation.

A few sperm cells manage to reach an oviduct.
Many sperm cells are released from the penis.
Here it settles into the endometrium.
Only one sperm cell fertilises the egg.
Then they swim up the uterus, into the oviduct and towards the egg.
As they swim some of them go the wrong way and some die.
The fertilised egg travels back down to the uterus.

____
____
____
____
____
____
____

6. (a) The placenta is formed from tissues of the _______________ and of the _______________

(b) Name two substances that pass across the placenta from the mother to the foetus:

(i) _______________

(ii) _______________

(c) By what method do substances pass across the placenta?

____

(d) State a function of the placenta other than the exchange of substances.

____

(e) Name one harmful drug that may pass across the placenta.

____

(f) Name one harmful virus that may pass across the placenta.

____

(g) What happens to the placenta following the birth of the baby?

____

7. (a) Where is oestrogen produced? _______________

(b) Give one function of FSH in:

(i) Males. _______________

(ii) Females. _______________

(c) Where is progesterone active? ______________________________

(d) Name a hormone that inhibits FSH. ______________________________

(e) Where is LH produced? ______________________________

(f) If pregnancy occurs name two hormones that are involved.

____

SECTION C

1. (a) Explain each of the following terms: ovulation (R), fertilisation (S), implantation (T), placenta formation (U). Draw a diagram of the female reproductive organs and, using the letters indicated, mark the position of each event on the diagram.

(b) At the end of a pregnancy the baby is born. Outline the process of birth in humans.

2. A husband and wife, who are not using any method of contraception and are having intercourse on a regular basis, find they are not able to have a baby. List four possible reasons why they are unable to conceive.

3. The graph shows the changes in hormone levels during pregnancy (Fig. 33.3).

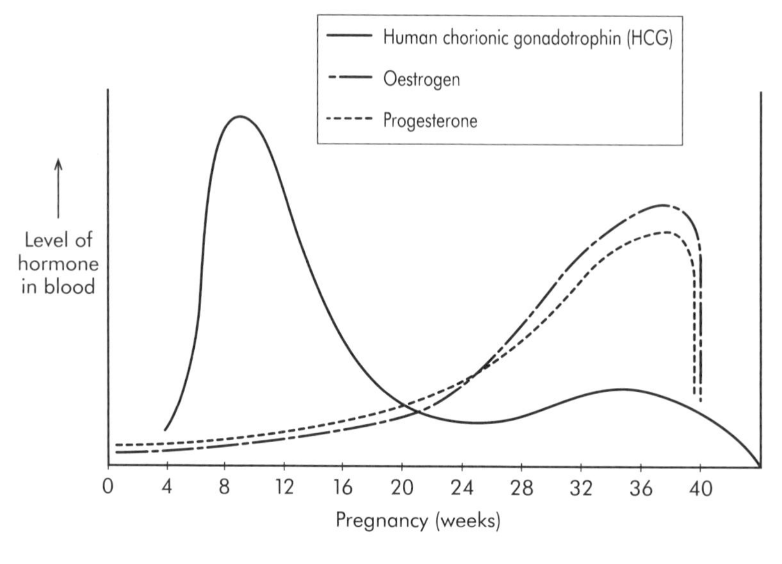

Fig. 33.3

(a) What structure produces HCG?

(b) Outline the changes taking place in the levels of HCG during the time shown in the diagram.

(c) At what time in the pregnancy does the level of HCG decline? What is the reason for this decline?

(d) Progesterone levels remain high during pregnancy. What are the two sources of progesterone?

(e) What effect does the drop in progesterone level at the end of pregnancy have on the uterus?

(f) During pregnancy the breasts are stimulated to develop milk ducts. Following the birth of the baby these milk ducts begin to secrete milk. Name the hormone involved in the production of milk and outline the benefits of breast milk to the baby.

4. (a) If a female cycle is 33 days, on what days in the cycle could intercourse lead to conception?

(b) Outline the development of the human fertilised egg up to the formation of the blastocyst.

(c) How does the blastocyst obtain its nutrition?

(d) How many chromosomes are there in a human fertilised egg?

(e) What name is given to the process whereby the blastocyst embeds itself into the endometrium?

5. The diagram shows the human embryo at 6 weeks (Fig. 33.4).

(a) Name A and B.

(b) What is the function of B?

(c) Which of the following tissues is the first to be formed in the developing embryo: the lungs, the ears, the nervous system, the bones?

(d) All of the major body organs are formed by the time the embryo is 8 weeks old. What name is now given to the embryo? Why would the embryo not survive if it were born at this stage?

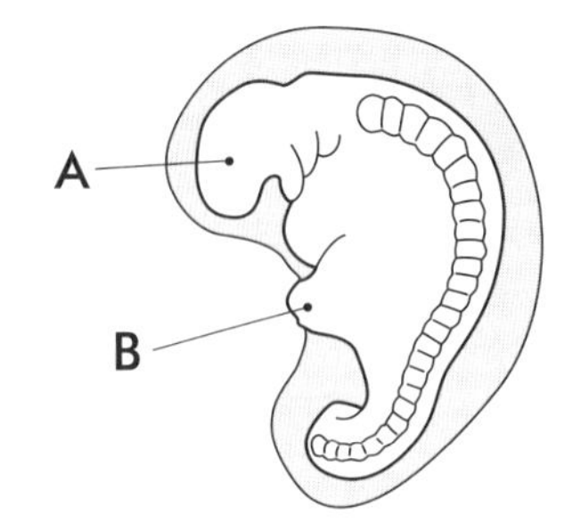

Fig. 33.4

6. Contraception may be defined as any means that prevents conception occurring.

(a) Distinguish between natural and mechanical methods of contraception, giving examples in each case.

(b) Answer the questions below by selecting from the list of contraceptive methods which follow:

the pill the condom the diaphragm
the rhythm method a vasectomy

(i) Which method(s) prevents eggs being produced?

(ii) Which method(s) is used by the man?

(iii) Which method(s) also gives protection against some sexually transmitted diseases such as AIDS?

(iv) Which method(s) prevents the sperm reaching the egg?

(v) Which method(s) depends upon knowing the woman's fertile period?

7. The diagram shows the position of a baby just before it is born (Fig. 33.5).

(a) Name the structure labelled X.

(b) What fills the cavity (space) labelled Y?

(c) Name one substance that passes to the baby from the mother through structure Z.

(d) Name one substance that passes from the baby to the mother through structure Z.

(e) The baby began life as a result of fertilisation. Copy the table below and complete it by naming the two cells which joined in fertilisation and state where each cell was produced in the body.

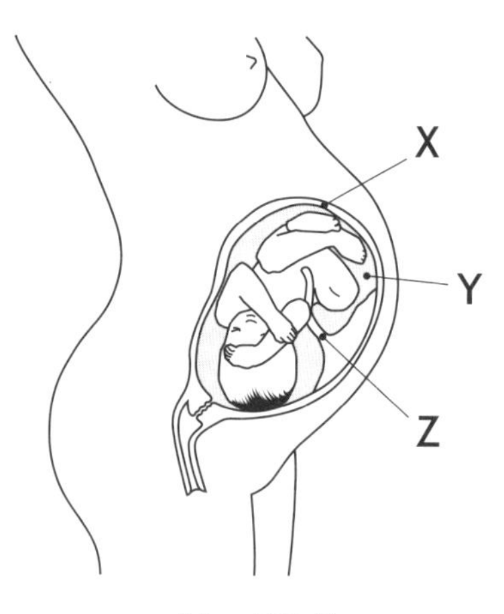

Fig. 33.5

Cell	Where cell was produced in body
1.	
2.	